ICE AGE ... 2025

**Why it's happening.
Important steps to get the country and your family ready**

Dr. Joel H. Glass

© Copyright Delaware River Publishing, 2019
All Rights Reserved.   ISBN: 978-1099083662

Delaware River Publishing
10685-B Hazelhurst Dr.
Houston, TX

delawareriver@usa.com

TABLE OF CONTENTS

DEDICATION

ABOUT THE AUTHOR

CHAPTER 1

CARING GLOBAL WARMING JUSTICE WARRIORS KILL THE WALRUSES

"OUR PLANET" NETFLIX TV SERIES CREW APPARENTLY SPOOKED WALRUSES OVER

SOME CLIFFS, THEN BLAMED GLOBAL WARMING

CHAPTER 2
WHAT ABOUT ALL THOSE SCIENTISTS?

THE LIE ABOUT 97% OF SCIENTISTS...

CHAPTER 3
IS CARBON DIOXIDE (CO2) GOOD OR BAD?

CHAPTER 4
WHAT CAUSES A RISE IN GLOBAL TEMPERATURES WHEN IT OCCURS... AND DO PEOPLE HAVE ANYTHING TO DO WITH IT?

CHAPTER 5
WHAT WOULD IT MEAN, IF GLOBAL WARMING IS CAUSED BY CO2?

CHAPTER 6
BUT WHAT ABOUT ALL THOSE SCARY REPORTS BY THE UN AND UNIVERSITY SCIENTISTS ?

CHAPTER 7
MORE ABOUT THE CRAZY SCIENTISTS OF THE GLOBAL WARMING FRAUD

THEY SAY THEY ARE THOUSANDS BUT THEY DON'T HAVE ENOUGH TO FILL A BUS

CHAPTER 8
GLOBAL WARMING, AND GLOBAL COOLING SUCH AS ICE AGES, DO OCCUR.

BUT ARE THEY CAUSED BY HUMAN ACTIVITY?

CHAPTER 9
WHAT IS A "GREENHOUSE GAS" ?

CHAPTER 10
THE REALLY STRANGE THING ABOUT THE GREENHOUSE GASES COMPARED TO OTHER GASSES

AND THE AMAZING TRUTH HOW LITTLE CO2 IS AROUND

CHAPTER 11
SOME LITTLE NUMBERS TELL A BIG TRUTH

CHAPTER 12
BUT WHAT ABOUT THE GREEN NEW DEAL ?

WON'T ALTERNATIVE ENERGY LIKE WINDMILLS AND SOLAR PANELS HELP US ?

MAYBE THAT'S SOMETHING GOOD WHICH IS IN THE GLOBAL WARMING MOVEMENT ?

CHAPTER 13
THE NEXT ICE AGE.
COMING TO YOUR NEIGHBORHOOD VERY SOON.....   DON'T MISS IT!

CHAPTER 14
BUT WHAT ABOUT THE LATEST SCARY PAPER ON GLOBAL WARMING...

AS USUAL, BIG, REALLY BIG MATHEMATICAL AND CONCLUSION MISTAKES IN IT...

I WONDER WHY?

CHAPTER 15
THE NEW ICE AGE IS COMING SOON

DOES THE AUTHOR OF THIS BOOK ACTUALLY KNOW SOMETHING ABOUT COLD WEATHER?
WELL..... YES

CHAPTER 16
TOUCHY-FEELY PROOF A NEW ICE AGE IS COMING VERY SOON

CHAPTER 17
THE DEMOCRAT MEDIA COMPLEX:
BRAINWASHING THE COUNTRY ABOUT GLOBAL WARMING

KEEPING US FROM GETTING PREPARED FOR THE REAL DISASTER

CHAPTER 18
OUR PUSHBACK AGAINST THE DEMOCRAT MEDIA COMPLEX IN THE BATTLE FOR TRUTH

FREE FOR THE READERS OF THIS BOOK, OUR VERY OWN NEW AND AMAZING VIDEO

JUST GO TO THE YOUTUBE LINK, AND SEE THIS AMAZING VIDEO ABOUT GLOBAL WARMING FRAUD, UNLIKE ANY OTHER YOU HAVE SEEN BEFORE

CHAPTER 19
THE GLOBAL WARMING & CLIMATE CHANGE CAUSED BY HUMAN ACTIVITY FRAUDS HAVE ALREADY STARTED KILLING US

CHAPTER 20

WHAT ABOUT THE GLOBAL WARMING BILLIONAIRES?

CHAPTER 21
BIG GLOBAL WARMING MONEY IS VERY REAL!

IT'S A HUGE AMOUNT...

AND WE AS TAXPAYERS ARE PAYING FOR MOST OF IT

CHAPTER 22
GLOBAL WARMING DISASTERS... DREAMS OF THE POLITICAL LEFT THAT NEVER COME TRUE

WHAT TO DO ABOUT RISING SEA LEVELS?
RELAX, IT'S A FAKE LIKE EVERYTHING ELSE

WHAT ABOUT THE GOVERNMENT GLOBAL WARMING REPORT, SPRUNG AS A TRAP ON THE PRESIDENT IN 2018, WHICH FORTELLS DISASTER?
RELAX IT'S A FAKE TOO.

CHAPTER 23
WHAT CAN WE DO? IF ANYTHING, TO PREPARE AMERICA FOR THE COMING ICE AGE?

CHAPTER 24
SAD NEWS FOR GLOBAL WARMING RELIGIONISTS.
AND FOR THE REST OF US ALSO.

THAT THE NEW ICE AGE IS STARTING SOMETIME BETWEEN 2019 AND 2035. VERY LIKELY ABOUT 2025

SEVERAL MEASUREMENTS, THE OCEAN CURRENTS, SOLAR ACTIVITY, AND ATMOSPHERIC HEAT LOSS INTO SPACE INDICATE THE ICE AGE IS ON OUR DOORSTEP

CHAPTER 25
WHAT ABOUT SOCIAL JUSTICE?

CHAPTER 26

WHY AND WHEN THE NEW ICE AGE WILL START

CHAPTER 27

HOW TO GET THE COUNTRY READY... HOW TO GET YOUR FAMILY READY FOR THE SEVERE ICE AGE JUST DOWN THE ROAD

CHAPTER 28
THE SURPRISING NEWS HIDDEN BY THE DEMOCRAT MEDIA COMPLEX:

A REVOLT BY THE PEOPLE AGAINST NATIONAL GOVERNMENTS AND MULTINATIONAL GLOBAL WARMING POLICIES IS TAKING PLACE ACROSS THE WORLD.

EVEN IN SOME STATES IN THE USA IT'S STARTED

LET'S ALL AMERICA REVOLT AGAINST GLOBAL WARMING FRAUD BY USING POLITICAL, JUDICIAL, AND MEDIA EFFORTS

CHAPTER 29
THE ICE AGE START IS NIPPING AT OUR TOES

CHAPTER 30
THIS IS A BONUS CHAPTER FOR THE READERS OF THIS BOOK

IT'S ABOUT THE NEW GLOBAL WARMING BOARD GAME

CHAPTER 31
A POLITICAL WAR AGAINST WE THE PEOPLE, HAS BEGUN. STARTED BY THE CLIMATE CHANGE POLITICAL LEFT. AND MAKE NO MISTAKE ABOUT **IT, THEY** HATE US.

IT'S ON. THE POLITICAL, JUDICIAL AND MEDIA ICE AGE WAR IS DECLARED

THE ICE AGE WILL BE HERE BETWEEN 2019 AND 2035 (around 2025), AND THE GLOBAL WARMING MOB IS HOLDING US BACK FROM PREPARING FOR IT

CHAPTER 32
IS THE GLOBAL WARMING MOVEMENT REALLY SATANIC ?
LET'S USE THE RED CHECK LIST AND FIND OUT

CHAPTER 33
WHAT CAN WE DO NOW, AFTER READING OUR NEW BOOK
CAN WE STOP THE COMING ICE AGE?
GETTING READY FOR IT... A SUMMARY

CHAPTER 34
THE SAD BUT PREDICABLE GENOCIDAL NATURE OF THE GLOBAL WARMING /
CLIMATE CHANGE FROM HUMAN ACTIVITY MOVEMENT

AND WE BRING A CHALLENGE TO THEM

CHAPTER 35
THE GREEN NEW DEAL...

AND THE PEOPLE BEHIND IT, ARE A THREAT TO NATIONAL SECURITY OF THE UNITED STATES

AND THIS IS NOT A JOKE. THEY REALLY ARE

WHO THEY ARE AND WHAT THEY ARE DOING

CHAPTER 36
COUNTDOWN

WHEN THE ICE AGE IS GOING TO BEGIN... AND HOW YOU CAN SAVE THE SEA OTTERS

AMAZING GRACE

DISCLAIMER

DEDICATION

I thank my Mom and Dad who encouraged me to learn, to tell the truth. And who provided great role models for me.

Also my wife and daughter. My wife for great ideas and for keeping me from bouncing around like a ping-pong ball on this book and other work. And for being encouraging in times of challenges.

My daughter of five, thanks for an amazingly funny sense of humor. Her hero is Winston Churchill.

Also thanks to my scientific colleagues, Dr. Jay Lehr and many others, leading scientists in the US. They have encouraged and supported this book.

And now, dear reader, we're about to enter the upside down world of Global Warming which is sending the country in the wrong direction.

Instead of getting ready for warming, and by focusing on CO2, which as you will see has no effect on climate... we should be hardening our infrastructure for a new Ice Age. So buckle up and hold on.

ABOUT THE AUTHOR

Dr. Joel Glass is an engineer working many years in alternative energy.

He has worked on alternative energy project teams for hydropower, wind power, solar thermal, and solar photovoltaic systems. These projects in the US, Canada, Malaysia, South Korea, China, and the EU (Scandinavia).

He is a specialist in cold weather climatology. The causes and effects of extreme natural cold weather on infrastructure and life... having worked for long periods of time in very very cold places; including the northernmost inhabited village in the world. This work has helped him understand the nature of the coming Ice Age. And explain what can be done to prepare for it.

He is an accredited member of the press. A conservative, constitution-supporting conservative Republican.

This book is written for all Americans. It is written under the First Amendment.. Freedom of the Press and Freedom of Speech.

A supporter of President Trump, who is completely correct on climate science and climate policy issues.

When the author approaches writing this or any other book, he goes for the truth, and writes it. Let the chips fall where they may.

What he discovered in two years of research, is that the truth supports the position of President Trump: Global Warming / Climate Change Caused by Human Activity is not only fake news, it's fake science.

And that the Global Warming / Climate Change From Human Activity movement is not actually related to climate. It has nothing to do with climate.

It's related to money and power. It's a power grab by the political left basically to remove the basic rights of the citizens of the United States. It's a totalitarian movement.

INTRODUCTION

IS AN ICE AGE COMING SOON.

AND ALONG WITH IT THE COLLAPSE OF THE DEMOCRAT PARTY... WHICH HAS BECOME A TOTALITARIAN ORGANIZATION BASED ON NAZI POLITICAL AND PROPAGANDA METHODS

Dear reader, by or about the year 2025 two significant changes will have occurred.

These will change human life on earth forever.

First, the new Ice Age will be underway. If it's not the full force downward plunge of temperatures, it will be at least that we have gone over the cliff. The latest date for this new Ice Age plunge will be 2035.

Second, the Democrat Party of the United State, along with it's relatively newly spawned vipers, the Justice Democrats / Democratic Socialists will have experienced two major completions of its recent development.

The political party will have become a full-blown Nazi – Soviet type totalitarian organization. Attempting to install an all powerful state. A state in which the individual does not exist. In which individual rights to not exist.

Along with that horrific change, will come the destruction of the Democrat Party and its vipers. This destruction will probably take a political, judicial, and media form. Their organizations will collapse. Or, and this is very possible, there will be a change by force. In which the will and might of the majority, which is how decisions are supposed to be made in the United States according to the constitution, will destroy them my any means necessary.

It means that the year 2025 is not going to be a particularly cheerful year.

However, the experience and nature of 2025 depends on what we do between today and 2025.

This book contains detailed information on human life, plant life, animal life during the last Ice Age. From 1300 to 1800 there was a taste of what we have now in front of us.

It's not a particularly good taste. And contains the spoiled foods of societal collapse, disease, civil war of neighbor against neighbor for survival. But if we prepare the water, food, and fuel infrastructure, and become ready in terms of training and education, none of these awful effects need take place.

As for the Democrat Party, there is no preparing, no changing them for the better. They are ideologues, brainwashed mind-numbed robots. For them, 2025 is the end. Before that, however they will have morphed full blown into the Nazi political power and propaganda organization described in detail in this book.

But isn't this book about climate? Isn't this book about the coming Ice Age?

Yes it is, and most of the book deals with the coming 2025 Ice Age and how to prepare for it. Both nationally and for your family.

The coming Ice Age is inseparable from political life, and they effect each other.

Standing between chaos, mass death, starvation, violence of an Ice Age society, is the Republican Party. The Party of Abraham Lincoln. The Party of Ronald Reagan. And the Party of Donald Trump. The Party of Life, Liberty, and Happiness. The Party of the Future.

HOW THIS BOOK STARTED, AND WHAT WE DISCOVERED

This book started as a guide out of darkness for those on the political Left, who believe in Global Warming.

And it is also written for Republicans, conservatives or anyone who has relatives, friends, co-workers who are Leftists and who believe in Global Warming, so you can explain things to them. And bring them back into reality.

It's a book for people who truly are environmentalists. Who know or will know that Global Warming and Climate Change – Caused - by human activity, is a gigantic fraud. We can save the polar bears, sea otters and a lot of people by understanding the inconvenient truth.

That inconvenient truth is that it's going to get a whole lot colder, not hotter. This will happen, and very soon, not because of human activity. But it will happen because the sun is entering the kind of low activity phase that causes ice ages on earth.

There is nothing we can do about it, except to understand the challenges and to prepare for it.

As I wrote this book, I realized that the false belief in Global Warming Caused By Human Behavior is more than just wrong. It's dangerous. Deadly dangerous. And I will explain why.

First, the Global Warming alarmists, are trying to politicize and even to criminalize anyone who has science and facts to refute the false religion they believe in.

Billionaire Tom Steyer suggested, seriously, that President Trump be impeached because he has the "facts" wrong on Global Warming and does not embrace the idea that human activity is heating the planet.

Mr. Steyer is not alone. Other billionaires fueling the lie include Bezos, Soros, Bloomberg and others. They bankroll an attempt to crush both scientific evidence and crush those who have already seen the light about the Global Warming fraud..

The motivation is political. It's simply an attempt to utilize the BIG LIE as a way for Democrat Party, Justice Socialists, to destroy the rights of Americans enumerated in the Declaration of Independence and the Constitution.

The politicization and even efforts to criminalize the scientific point of view that Global Warming theory is inaccurate, is both troubling and dangerous for the country and its citizens.

In 2009 I wrote a seventeen page paper, which clearly presented world leading climatology-related scientists making the point that Global Warming is the largest scientific and financial fraud in world history.

I presented all the necessary data, to show that this is the case. And I ended the paper saying that far from Global Warming, the earth would be entering a new ice age in about 15 years. Because temperature on earth depends on.... the sun. And the sun is entering a very cool phase called a Solar Minimum.

A weak thermal output by the sun dramatically cools the temperatures on the surface of the earth where we all live. Even at that time (2009) it was clear that we were on the way to entering into a solar minimum. The sun is becoming relatively inactive.

Regarding the coming new ice age, I regret to inform those of you on the political Left of some unpleasant news:

The New Ice Age is upon us. Because the sun is cooling down very very strongly. It happens in cycles and we are entering a cooling cycle now.

This isn't some apocalyptic view based on writings of Nostradamus 463 years ago. It's based on the most recent analysis of solar activity by key governmental and academic solar research, including NASA. And we here on earth are in for a very very dangerous ride.

An ice age is not pleasant. And we are ill prepared for it. We are unprepared in main part because the unhinged politically motivated Global Warming hysteria has sent us off in the wrong direction.

The next ice age could begin any time soon. Now or at latest around 2035. Best guess for me being around 2025.

It's always difficult to pinpoint exactly when such solar events will begin but all the evidence recently in solar activity, atmospheric heat loss, and ocen current data, point to an ice age hitting us hard very very soon.
.
This book has an additional purpose to informing the citizenry what is happening. It's a call to climate arms. To alert readers to the catastrophic effects which even Small Ice Ages bring.

Several chapters discuss in colorful terms the ICE AGE, explaining in details how that will effect us.

The story isn't pretty. But forewarned is forearmed.

Now in accordance with my original plans, and

- so that this book can actually help people on the political Left who have been lied to, to understand things

- and to help Republicans use this as a tool to help rescue their Leftist friends and relatives, who are drowning in a sea of lies

This exciting book, is in a sense like mystery novel. We are going to find out what sins and perhaps actions which could be subjects for criminal referral... investigated to determine possible criminal intent, which the Democrat Party, the Global Warming Mob, and the Democrat Media Complex are motivated by. It's about money. It's about power.

And to do so the book presents things in a very very clear way. And in straight-forward language.
Based on that clarity, even people who have been brainwashed and who are part of the Global Warming false religion, can understand what actually influences the climate of our world. And what is coming soon down the road in terms of climate. And here's a hint, get some long underwear.

While those of the political Left, who are engaged in the greatest financial and scientific fraud in history, will not likely change... the book can serve for the rest of us as a handbook to get ready for the coming Ice Age.

- First to harden the US water, transportation, housing and food infrastructure so it can withstand the coming unusually low temperatures.

- Second, to personally get ready as best as we can (I'm not a gung-ho survivalist, but some understanding of ultra-cold weather will be necessary) to make it through even a Small Ice Age

- to work as groups, organizations, some as part of the Republican party, and with our Republican representatives in government, to secure rational legislation and policies that assist in getting country ready for the coming new Ice Age catastrophe

This could include (but not be limited to) doing three things:

- VOTE REPUBLICAN, as if your life depends on it. Because it does. And vote for constitutional, conservative, aggressive candidates who support the president.

    RINO (Republican In Name Only) representatives such as Mitt Romney, will have to retire to a dumpster and leave the government to sane, patriotic people.

    Fortunately we have a president is right on Global Warming / Climate Change. Let's support his efforts.

- Crush any type of Democrat Party, Justice Democrat voter fraud and illegal alien voting.

    Use masses of trained election volunteers and Republican party attorneys before and during the election process itself to stop the Democrat theft of elections.

    Democrat Party, Justice Democrat voting fraud not only destroys the democrat election process, but in addition if we continue heading along the Global Warming Avenue of Lies, as a nation we will perish. They are sending us in the wrong direction.

- Crush the unhinged totalitarian efforts of the Democrat Party, Justice Democrats in congress to ramp up programs for fighting non-existent Global Warming.

    These insane programs won't serve any useful purpose and will destroy the American energy infrastructure. No more heat, no more electricity.

    These irrational plans, such as the Green New Deal (which is actually a power-grab and is the Green New Steal), the Democrat Party, Justice Democrats, its street mob, and the Democrat Media Complex are pushing the country into national suicide.

- **I will repeat that: The Democrat Party is pushing the United States into national suicide.**

    And that We the People do not like it, should be made very clear to them by crushing them in elections and in the media.

- Identify those who are pushing the fake Global Warming agenda. And hold them responsible for trying to send us into national suicide. We can protest and utilize political campaigns, judicial efforts and media campaigns in a more effective way than they can.

To make it clear what we are up against when the new Ice Age shows up, it's going to be so cold during the winters for at least three hundred years (and maybe a lot more if it's a major ice age and not a Little Ice Age)... that about half or more of the political Left bastion New York City, will die from the effects of the freezing cold during the first few years.

To those of you on who think you are on the political Left, it's important to you, to your happiness, that you stop playing games with the future of your family, yourself, and your country. Slow down and think.

Right now you have been brainwashed to believe in something this is simply a lie. It's not only not true, but it is a purposeful lie, created by a few so-called scientists and those with financial interest who want to achieve political ends and get a pile of money by lying.

CLIMATE CHANGE / GLOBAL WARMING FROM HUMAN ACTIVITY IS... THE BIG LIE

As we will read in this exciting book, the climate movement of the Democrat Party, Justice Democrats is itself THE BIG LIE. And we will read and see that it is no different in political method and propaganda method than the BIG LIE of the Nazi movement of Germany or the Stalinist government of Russia.

Global Warming people, you believe in a lie. And if you are chasing a lie, the results can not be good for you, for your family, and your future.

And this lie you believe in, Global Warming and Climate Change caused by human activity, has wrongly influenced the allocation of American resources to get ready for the future. We are preparing for a problem that does not exist: Global Warming.

If the country does not act to prepare for the coming Ice Age, we are all in serious trouble.

So read this book carefully. And as a great way to complete the introduction, let's hear from three great climatologists:

"Warming fears are the "worst scientific scandal in the history… When people come to know what the truth is, they will feel deceived by science and scientists."

UNIPCC (United Nationiona Intergovernmental Panel on Climate Change) Dr. Kiminori Itoh, award winning PHD environmental Chemist

AND

"CO2 emissions make absolutely no difference one way or another…. Every scientist knows this, but it doesn't pay to say so…Global warming, as a political vehicle, keeps Europeans in the driver's seat and developing nations walking barefoot."

Dr. Takeda Kunihiko, vice-chancellor of the Institute of Science and Technology Research at Chubu University, Japan.

AND

The hysteria about Global Warming and Climate Change "is not only fake news. It's fake science."

Pat Moore, Co-Founder of Greenpeace environmental organization, Board of Directors, International Climate Science Coalition

CHAPTER 1

CARING GLOBAL WARMING JUSTICE WARRIORS KILL THE WALRUSES

'Our Planet' TV Series Crew Apparently Spooked Walruses Over Some Cliffs, Then Blamed Global Warming

There has been a major debate about a center and keystone a new nature televison series "Our Planet".

The series, ordered by the fund-raising billionaires at the World Wildlife Foundation, supposedly is to help us see our beautiful wildlife of earth. In reality it's a propaganda series to promote the Global Warming fraud.

The main idea of the series, is that the present Global Warming (which is in reality non-existent) is driving animals not only to extinction, but also to suicide.

Scenes in which a group of walruses jumped off a cliff, supposedly in an act of suicide.

(Happy walruses, before the "Our Planet" film crew invaded their space and caused them to panic and stampede)

And isn't it amazing that the producer, director, film team of "Our Planet" just happened to be in the right place at the right time to see it?

Similar to CNN being at the right place and exactly at the right time, when Mueller's FBI arrested Roger stone. The FBI showed up at 7:00 AM and CNN set up shop outside his house at 6:45. Simply amazing.

And regarding the walrus "suicide," the "Our Planet" film crew got it in living color. It became a major marketing focus for their programs and Global Warming informational effort.

The "Our Planet" series is, however, not so much a nature series, as a propaganda effort to ramp up the hysteria.

This Global Warming propaganda objective of the producers and directors is obvious since they readily admitted it in a nationally televised interview on PBS. One led by interviewer Christine Amanpour and having the directors and producers of the "Our Planet" series.

Anyway, the "Our Planet" crew "just happened" to be at the right place at the right time, as the walruses decided to commit suicide over Global Warming, and jump from cliffs.

POLAR BEARS?

Experts in walrus behavior have responded to the absurd notion that the poor things were killing themselves because of despondency over Global Warming. Those experts informed us, that this behavior of jumping off cliffs is not out of the ordinary in certain circumstances.

And that it occurs, for example, when a polar bear or several polar bears show up and panic the walrus herd. Which then heads blindly and in a panic over a cliff that may lead away from the oncoming threat.

Well, that sounds like a more reasonable explanation than depression over non-existent global warming, doesn't it?

Yet the film crew, producers/directors, key members of them being present for the walrus stampede, have denied that there were any polar bears in the area.

And you know what, they probably were right.

From their nasty video drones, which take the footage, and send back live images to them, they could have spotted any polar bears in the area. And even got it on tape.

IF NO POLAR BEARS, THEN WHAT?

An interesting bit of information came my way recently. It's not only polar bears that panic walruses.

It can be anything out of the ordinary in the walrus environment. Anything startling that they are not used to. That perhaps they have never encountered, which can send them into a panicked herd frenzy.

And one of the things mentioned by wildlife experts is that when an airplane comes more or less out of nowhere and flies over them, it triggers their panic response.

Well... the film crew didn't have an airplane... but they had drones.

And the drones can fly very very low relative to the walruses. And of course, did so in order to get their video footage.

What apparently happened, is that the wildlife geniuses and experts of "Our Planet", managed with their drones to drive the walruses into a panic. Right over the cliff.

THE "OUR PLANET" TEAM PROBABLY REALIZED IT, AFTER IT HAPPENED

The drone or drones go up and the walruses go over the cliff. This is fairly straightforward.

But yet, Global Warming got blamed by the drone masters.

It's bad enough the series' experts and film crew apparently killed the walruses through negligence.

But to add insult to injury, they then blamed it on "Global Warming." As the political Left always says: "Never let a good disaster go to waste."

THE LESSON: NEVER TRUST A GLOBAL WARMING / CLIMATE CHANGE-CAUSED-BY-HUMAN-BEHAVIOR PERSON

In the case of "Our Planet"... **The Masterminds Behind Netflix's 'Our Planet' Admit The TV Series Is Intentionally Alarmist**

If there was any doubt that the Netflix paid-for series was purely a propaganda effort of the global warming movement, then that doubt has been dispelled after two of the masterminds behind it, a producer and a director, admit as much in a PBS television interview.

In an exchange led by the ever-reliable Left-wing political hack Christine Amanpour, the show's masterminds, Kieth Scholey and Sophie Lanfear, bluntly admit the purpose of the series was to generate alarm about how "**global warming is destroying the natural world.**"

You can see this amazing spilling of the beans at PBS.

That they would admit that the so-called "nature series" is actually a propaganda series is a bit bizarre.

It's climate hysteria in full bloom. Producer Kieth Scholey stated it's a nature series, but we have to do something now [about non-existent] global warming or we will lose that nature, and even the whole biosphere! YIKES.

So it's really just a propaganda effort. And they blurted it out, simply because they are delusional.

They played a clip about the unfortunate walruses. Walruses that they evidently killed by using their drones and causing the herd to panic and stampede over the cliff.

But the cliff jumping was presented as occurring because the walruses were depressed about global warming.

How they could keep a straight face when covering up their apparent negligence and walrus killing is unbelievable.

This one example is why I believe my language in this book that the Climate Change From Human Activity Movement is "rotten to the core", is quite accurate.

The motivation of Climate hysteria is money, or a political power grab, or a combination of both.

It not only allows, but motivates people to lie with impunity. The book is half about the technology of Global Warming and half about the totalitarian nature of the movement.

The hidden problem of the Climate hysteria movement, is this:

The United States, and your family, could be spending time and resources preparing national and personal infrastructure for the coming Ice Age. 2025.

But instead we are spending world wide more than 30 billion dollars yearly chasing a non-existent nightmare.. Global Warming / Climate fraud.

In any case, now regarding "Our Planet", what is apparently a despicable double act has come to light. First killing the walruses through negligence and stupidity, and then lying about it, as a cover-up.

CHAPTER 2

WHAT ABOUT ALL THOSE SCIENTISTS?

THE LIE ABOUT 97% OF SCIENTISTS...

Perhaps it's a good idea to deal with the biggest and most prevalent lie of all... that 97% of scientists support the Global Warming / Climate fraud.

Although it's just is one many lies of the political Left uses to falsify what is actually happening and to crush debate and truth, it's a real whopper.

In fact there no such large group of scientists that are pushing the Global Warming agenda. 97% is a fake number.

And the few "scientists" that are pushing the Global Warming fraud, are in it for the money. The book explains how they get huge amounts of research grant money in return for lies. And we are not talking about candy store change, but billions of dollars yearly courtesy of the American taxpayers and a few Left wing billionaires.

So where did this particular lie about the 97% of scientists come from? Like most of the so-called "facts" about Global Warming, it's a purposeful statistical and mathematical mistake.

A mistake made on purpose.

Now before your very eyes in this exciting book, the concrete curtains are pulled back and we see what is really going on.

In fact, the situation is like this: out of the hundreds of thousands of scientists in this country that work in a field which has some relationship to climate,

it isn't 97% of scientists that support the believe in Global Warming.

It's 97 scientists !

Less actually.

That's it. About 97 scientists support it, and they are lying about the science, for political purposes and for financial gain. Hundreds of thousands of scientists recognize that the idea of human-caused global warming and climate change is a fraud, a lie and is wrong.

REVEALED:

WHERE THE 97% LIE CAME FROM (AND HOW THE LYING GLOBAL WARMING MOVEMENT USE IT)

It's not 97%. It's not even 97 people. It's just 75 !

Just 75 of them agreed in a poll they supported Global Warming. 75.

And guess what, ten thousand scientists who were invited to take the poll, which was based on the idea that Global Warming exists. And overwhelmingly those scientists polled, showed by not participating in the poll that they did not believe Global Warming exists and is a result of human activity.

Here dear readers, we now have the pleasure to rip to shreds, once and for all the "97% of scientists" lie.

It's a huge lie. About as big as you can get. In fact the percentage of "scientists" that supported the idea of man-made Global warming in this infamous original University of Illinois poll **was not 97.00% but was 00.0073% of the total scientists polled !**

Less than one one-hundreth of 1% of scientists who were polled supported the theories of man-made Global Warming, Al Gore and the IPCC.

The lie was fabricated by University of Illinois climate activists like this...

- The poll was conducted by the University of Illinois on-line, with 10,257 scientists that were polled

- And by twisting the results of that poll, the Global Warming fraud scientists who always speak with a forked tongue, have claimed over and over and over, that 97% of scientists in the world support the false religion of Global Warming.

But while over 10,000 scientists were polled, only 77 who worked in Climatology answered the poll. The response level was so low, because the poll was based on a false premise question that Global Warming exists:

THE TWO QUESTIONS IN THE UNIVERSITY OF ILLINOIS POLL WERE BASICALLY LIKE THIS.

GLOBAL WARMING EXISTS....

AND IT IS CAUSED BY HUMAN BEHAVIOR.

To give a positive answer you have to agree to both of these.

And of the 77 climatologists who answered the poll, 75 stated they did support Al Gore and the Global Warming idea.

This amazing lie, that 97% of scientists support global warming and that it is caused by humans, was based on just 75 people out of 10,000 who were polled.

And that's 75 out of 10,257. Almost nobody.

The tiny number of 75 "scientists" support Global Warming, frightened the University of Illinois climate science hysterics who had an agenda to give Al Gore a gift.

That only 75 "climate scientists" agreed with Gore out of 10,257 set off the alarm bell among Global Warming hoaxers. The poll showed that basically nobody agreed Global Warming Exists and also that it is Caused by human behavior.

So the pollsters at the University of Illinois decided to tell a little white lie. And in their dark moldy hallways, and small dark rooms without light and air, the university academic Global Warming social justice vampires twisted the results.

97% agree ! The number was trumpeted loud and strong world-wide.

And yes 75 people out of 77 climatologists who responded is indeed 97%. But somehow they forgot about the other 10,120 scientists who didn't agree.

- The poll itself was a manipulative effort by a few individuals at the University of Illinois to silence critics of the Global Warming hoax, before these critics they alert the people of America to the truth.

And additionally, and astonishingly, the poll was constructed by the University of Illinois climate activists, to leave out the key types of scientists who would certainly disagree with Global Warming, out of the poll altogether.

These branches of science excluded by the fake University of Illinois "97%" fraud poll were:

* **solar scientists**
* **astronomers**
* **meteorologists**
* **space scientists**
* **physicists**
* **and cosmologists**

These key scientists were ignored. Ignored because the vampires in the dark University of Illinois chambers, knew they would vote "NO!"

Millions of children and teenagers under age of 18 have been brainwashed about this 97% lie.

In the spotlight we have of course the "Sunrise" radical environmentalist movement children, whose parents have turned them into mind-numbed robots. Children who are not only spouting the lies they have learned like crazy parrots, but who are rude and arrogant to boot. They have learned well from the Global Warming hysterics paid to misinform the nation.

As huge a deal as it is, the "97%" hoax was not the first BIG LIE from "climate authorities" about the number of scientists in their camp of deception.

- Previously, after the United Nations IPCC report of 2007, it was announced that 2,500 scientists supported the UN Global Warming hysteria.

  2,500 scientists! Then the hysteria must have a factual basis. Global Warming must be true!

  Later it was confirmed that there was a teeny weeny

little mistake in the number of supporting scientists.

It was NOT 2,500 SCIENTISTS, BUT WAS 25 !!

If anyone told such lies under oath, they would be indicted, convicted and imprisoned.

Meanwhile, as mentioned elsewhere in this book, well over 50,000 scientists working directly in climatology or in other scientific fields directly related to climate issues, have signed statements that they find Global Warming / Climate Change From Human Behavior to be a lie.

In addition to this massive number of scientists, scientific organizations and groups have added their voice that Global Warming is a fraud, and a long list of these world-leading scientists and organizations is found towards the end of this book. And many of the scientists are specifically quoted in this book.

Here's what one of the real scientists has to say about the 97% fraud. And because the political Left loves big government, they must love this scientist because he is part of it:

"It is a blatant lie put forth in the media that makes it seem there is only a fringe of scientists who don't buy into anthropogenic (human caused) **global warming."**

- U.S Government Atmospheric Scientist Stanley B. Goldenberg

In fact as we will see in this exciting book, Global Warming and Climate Change Caused by Human Behavior is debunked by scientists who are not bought and paid for, or who are power-grab maniacs.

Now let's look at the teeny weeny tiny little group of pseudo scientists who lie purposefully to push the political power grab agenda of Global Warming and Human Caused Climate change.

THE GLOBAL WARMING / CLIMATE CHANGE MOB HAS A TRUTH PROBLEM.

IT'S IN PART A MATH PROBLEM THAT EVEN HIGH SCHOOL KIDS

WOULDN'T HAVE.

https://wattsupwiththat.com/2012/07/18/about-that-overwhelming-98-number-of-scientists-consensus/

The following chart explains it clearly:

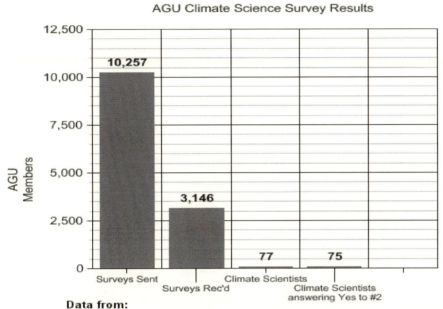

This is a great chart to reprint and put on your wall or desk. Or on billboards across the nation. From sea to shining sea. Let's give the truth some light.

And next time dear readers, any of your brainwashed friends, relatives, colleagues, media people use the 97% number, you know what to say. It's less than 00.0073%.

And you might as well mention to them at the same time, that the United Nations 2007 number of 2,500 Global Warming supporting scientists was admitted to be only 25 scientists. A teeny weeny little math problem. Oh well.

Put bluntly, Global Warming / Climate Change is a lie from beginning to its disastrous end. Don't let it ruin the life of your family, your country.

THE CLIMATE CHANGE MOVEMENT IS NOT ABOUT CLIMATE

Constitutional conservative republicans supporting the President (he is right on the Global Warming issue) must unite for truth. So we do not allow the political left and their Democrat Media Complex to bring a Soviet-style totalitarian government in this country by using a manufactured Global Warming crisis. To destroy the rights we have as We the People under the constitution.

We are going to need those rights when the coming Ice Age hits about 2025. Society will degenerate, and the only thing that can provide survival are the rights of the Declaration of Independence an the constitution.

"Climate Change" is not about climate.

It's about grabbing money and power.

And while the miscreants and reprobates of the Global Warming fraud, are grabbing money and political power... the nation is heading over the cliff of a coming Ice Age without being prepared.

A taste of the "new normal" is happening how. Today On November 20, 2018 according to wind chill factor data, the whole country is freezing. We dare any of the Global Warming mob to go and protest for a few hours in such temperatures. Go on, we dare you.

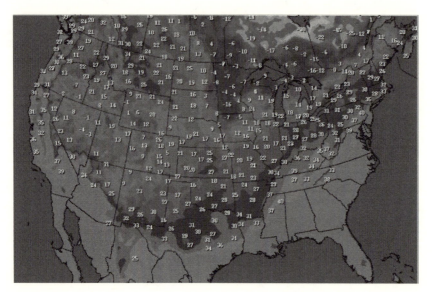

The earth is in the process of entering a cooling period (called a Solar Minimum). It may not have come in full force yet yet, but as you will read in this book, key scientific signs clearly show that the thermostat for the entire earth will be turned down in the near future. Welcome to the new Ice Age.

And now buckle up. We're going to see in Chapter 3 if Carbon Dioxide (CO2), is good or bad..

Here's a hint. We all exhale CO2. And plants love it. And we are going to need a lot of it after the new Ice Age hits, to help maintain what little agriculture will be possible.

CHAPTER 3:

IS CARBON DIOXIDE (CO2) GOOD OR BAD?

First of all is some news which will make those of you on the political Left very happy.

Periods of rising temperatures do indeed occur at times in the earth warming and cooling cycles. Or to be more accurate, climate period temperatures rise and fall with solar warming and cooling cycles.

But here is the bad news for those on the political Left – When periods of warming do occur, it's not because of anything human beings do, and not at all related to CO2 (carbon dioxide). It happens because of increased solar activity.

A leader of the Global Warming movement, Al Frostbite Gore, tells us that carbon dioxide is bad. Well, he says it's even worse than that, very very very bad.

Carbon dioxide is the very same gas used in soft drinks like CocaCola, Pepsi and all those energy drinks that those on the Left love, and use to keep moving forward through their swamp of lies.

So if Carbon Dioxide contributed to global warming, then anyone drinking a Coke or Pepsi or an energy drink, is destroying the planet. That's important to mention because these drinks are beloved by the political Left. Just go to any of the few and sparsely attended Global Warming demonstrations. Afterwards CocaCola, Pepsi, and energy drink cans litter the area.

Well, as good news to the companies that produce Coke and Pepsi, I should mention, that CO2 has no effect at all on climate.

The effect of these drinks on climate is zero. To the dear readers on the Left who are lucky enough to have a copy of this book, you can keep drinking and not feel guilty.

Our conservative constitutional republican readers, never felt the soft drink guilt in the first place.

CO2 AND CLIMATE:

Of course it's shocking to those of you on the Left, that the relation of CO2 to temperature on the earth is zero. No connection what-so-ever.

Many periods of the earth's climate were very cold, ice ages. And there was a very high CO_2 level, much higher than today. Today the CO_2 level is about 1,420 parts per million. And during very cold ice sages is was a few thousand parts per million.

Today the CO_2 level is actually very low in terms of earth history.

As mentioned today CO_2 atmospheric levels are about 420 parts per million. And during that time the earth was flourishing with beautiful plants an animals everywhere. Happy as can be. It was 1,200 ppm.

Also in earth's history, there have been very hot periods of temperature, in which the CO_2 level was low.

So that's the first thing. Global Warming does occur at times, but it's part of the natural cycle of the sun heating and cooling. And it's not related to human activity or CO_2.

What then exactly and in detail causes Global Warming? We now will take a look at that in Chapter 4. For now we know it's not CO_2.

If indeed CO_2 were to cause Global Warming, then we here on earth should work day and night to produce as much CO_2 as possible.

If it would cause warming, which it doesn't, then anything to help us make it through the coming 2025 Ice Age would be a welcome thing.

The new Ice Age, which is coming soon after the last one, which occurred from 1300 to 1800, should be arriving about 2025. And this book explains in detail how we know that. Thank you NASA. Thank you oceanographic leading professors. Thank you solar scientists.

CHAPTER 4

**WHAT CAUSES A RISE IN GLOBAL TEMPERATURES WHEN IT OCCURS...
AND DO PEOPLE HAVE ANYTHING TO DO WITH IT?**

Since Al Gore proclaimed in 2006 the imminent destruction of the earth due to overheating within ten years, there has been no significant rise in temperatures at all.

Any reported rise was a scientific hoax, using false data and manipulating data. And has now been corrected. The temperature has in fact been steady for decades.

But if some warming would occur in terms of a long time period and significant rise in temperature, it is caused by the same thing that Causes Global Cooling. That cause is: the level of activity of the part of our solar system that creates heat....

The Sun.

When the sun is very active, in a hot solar cycle, we have Global Warming.

And when the sun is not very active, in a cool solar cycle, we have Global Cooling. Ice ages on earth have been caused, by low solar activity.

To make this clear to those who have become mind-numbed (due to no fault of their own) by the Global Warming propaganda machine on TV, here is a computerized photo showing how this works:

WHEN THE SUN IS IN A HOT CYCLE, THE EARTH GETS HOT

WHEN THE SUN IS IN A COOL CYCLE, THE EARTH GETS COOL

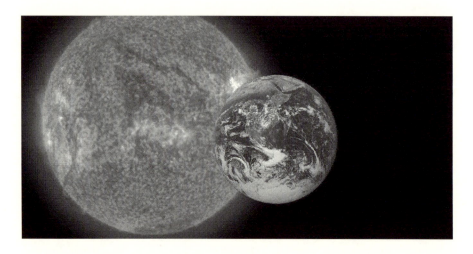

So, that's what causes Global Warming when it actually happens. The sun is in a hot cycle.

And that my friends, a hot cycle, is not happening now. Far from it.

We are heading full speed ahead into a new Ice Age caused by a cooler sun, a solar minimum. 2025.

CHAPTER 5

WHAT WOULD IT MEAN TO US, IF GLOBAL WARMING IS CAUSED BY CO2?

If global warming and Climate Change (all those nasty storms) were really caused by CO2, that would be very bad news for those of you on the Left.

Mainly because human beings exhale CO2. We are in fact, each and every one of us little CO2 factory.

Even Al Gore and AOC are CO2 factory.

And so are the puppet masters behind them: Zach Exley, Becky Bond, and several organizations financed by Leftist billionaires.

Therefore, if CO2 were a cause of Global Warming (which it is not, but let's just pretend that CO2 does cause Global Warming), there is a wonderful way for those who are strong advocates of Global Warming hysteria to help:

   Stop breathing

Yes. Those who are Global Warming / Climate Change protesters, you can take one for the team. Take a hit to save the planet. Stop breathing, and you reduce the CO2 level of the earth's atmosphere.

And even worse than that, you would have to get rid of carbonated beverages, like Coke and Pepsi, energy drinks and many mineral waters. Yikes.

But fortunately, CO2 has zero effect on the heating and cooling of the earth.

So nobody needs to "take one for the team".

Or even to give up their previous energy drinks.

Dear readers, not only is CO2 harmless for climate, CO2 is very good for living things of the earth.

It's part of the growth and life of plants. So all over the world, forests and crops are growing better because of CO2. And happy forests and sea plants, means happy animals and fishes.

So don't worry about it, and have lots of children. (Don't listen to AOC and her silly advice). Instead of not having children due to feat of imagined threats, have lots of new little CO2 factories. Your children. They are a blessing for life on earth.

But if the Global Warming hysterics desperately need something to worry about... Why don't they try worrying about the new Ice Age coming. Arriving around 2025.

CHAPTER 6

BUT WHAT ABOUT ALL THOSE SCARY REPORTS BY THE UN AND UNIVERSITY SCIENTISTS ?

If you don't want to read this chapter, we can summarize it on one line:

Assume that every academic and UN paper you read about Global Warming is a lie.

But of course it's more fun to read on and find out why they are lying.

Each year we are confronted with countless "academic" global warming publications which soon after publication are shown to contain data errors or misinterpretations.

Shown usually quite soon, to put it bluntly, to be frauds.

The authors are almost always tenured professors. And in one major case of 2018, there were 10 professorial Global Warming co-authors, and along with them was a "math" error.

This is the ten-author Resplandy et al. Ocean Heat Uptake paper, whose lead author is Dr. Laure Resplandy... a key member of the Princeton Department of Geosciences and Princeton Environmental Institute.

Very impressive of course. In fact, all ten authors have very impressive credentials.

The paper, **Quantification of ocean heat uptake from changes in atmospheric O2 and CO2 composition**, was published by the prestigious journal **Nature**.
https://www.nature.com/articles/s41586-018-0651-8

The title proved to be a bit embarrassing and ironic. For having the word "Quantification" in it. And the first word at that. "Quantification" is a long academic word that students in university learn. It means basically "counting". Adding, subtracting, dividing, things like that. And as it turns out, evidently the 10 authors of this paper, aren't able to do that.
Within a few days of publication of this paper, which was intended to be the

sledgehammer that destroyed any academic opposition to global warming hysteria, substantial math errors were spotted and became public.

This crashing down into the gutter did not take place before the New York Times, Washington Post, and the Democrat Media Complex had heralded this paper as **the end of "climate denial."**

It was apparently similar to the elation Hitler felt when visiting Paris after his military victory, and dancing a little jig in front of the Eiffel Tower before motion picture cameras. Instead of dancing a jig, the Democrat Media Complex sounded the trumpets:

"Startling!"

"This changes everything!"

"Global Warming is worse than we thought!"

"The end of Climate denial"

The Washington Post:
"The higher-than-expected amount of heat in the oceans means more heat is being retained within Earth's climate system each year, rather than escaping into space. In essence, more heat in the oceans signals that global warming is more advanced than scientists thought."

Yikes, ready the escape capsule !

Then shortly after publication, the findings of the paper were completely discredited (see the math problems here in a wonderful paper by brilliant mathematician Nicholas Lewis, a critic of Global Warming models: https://www.nicholaslewis.org/wp-content/uploads/2018/11/A-major-problem-with-the-Resplandy-et-al.-ocean-heat-uptake-paper.pdf)

After the truth became public, one of the ten authors Professor Ralph Keeling, a global warming proponent of a leading American Institute of Oceanography (located just as we say, spitting distance from my undergraduate college in California), made a public statement. Well they had to say something, didn't they.

To paraphrase Dr. Keeling's statement: "We seem to have made a teeny-weeny little math error. It wasn't very serious. But in fact, it did cause a little problem."

"But," he said (again paraphrased), **"What is important in science is not making mistakes. It's correcting your mistakes."**

The exact words were so thrilling, we might as well quote them:

"Our error margins are too big now to really weigh in on the precise amount of warming that's going on in the ocean…We really muffed the error margins."

"Unfortunately, we made mistakes here, I think the main lesson is that you work as fast as you can to fix mistakes when you find them."

Global warming support scientists are always the reincarnation of Goody Two Shoes. And a Goody Two Shoes knows what to do when they get caught fudging data, or downright lying in academic papers. Say, that he / she didn't have enough good data to "weigh in" in the problem. Oh my goodness…

And immediately, all was forgiven. The global warming paper that was hyped by the Washington Post, New York Times, and the whole Democrat Media Complex as the final proof that global warming exists, turned out to be scientific… ummm… well…. what some called "garbage". But that's OK because… being a Global Warming fraud scientist means never having to say you are sorry. Here we absolutely must insert an emoji to express everyone's happiness:

Sorry about our little mistake, said the authors. But now we fixed it and everything is OK.

Remember, there were ten global warming "scientists" who authored that paper. All professors or similarly academic "experts."

It's fair to assume that all of the "Gang of Ten" actually read this magnum opus they wrote and signed. And that they can do the basic math of a freshman college student.

In other words, it might be possible to succinctly paraphrase Professor Keeling's statement in this way: **"Sorry we got caught."**

Now, the "Computational" study is simply part of the unending global warming hysteria fudging academic data. In this case, wholesale falsification according to leading experts.

As we said, being a climate change supporting "scientist" means never having to say you're sorry. Except after getting caught. Then you have to seem a little sorry about something.

And as we bid farewell to this magnificent academic paper, let's remember who paid for this whole mess. It was us. We the People. The taxpayer.

Yes dear readers, without our knowing it, out hard earned money, we pay in taxes, financed this... uummm this 4 pages of "crap". The research was funded by the U.S. Climate Program Office of the National Oceanic and Atmospheric Administration. YIKES.

The difference between Global Warming fraud "scientists", and actual scientists who understand that the Global Warming / Climate Change Caused By Human Activity movement is based on lies, can be seen in the statement of the data true scientist fact checker, Nicholas Lewis, who originally found the error. He wrote as part of his clear beautiful mathematical analysis:

"On November 1st there was extensive coverage in the mainstream media[1] and online[2] of a paper just published in the prestigious journal Nature.

"The article,[3] by Laure Resplandy of Princeton University, Ralph Keeling of the Scripps Institute of Oceanography and eight other authors, used a novel method to estimate heat uptake by the ocean over the period 1991–2016 and came up with an atypically high value.

"The press release accompanying the Resplandy et al. paper was entitled "Earth's oceans have absorbed 60 percent more heat per year than previously thought", and said that this suggested that Earth is more sensitive to fossil-fuel emissions than previously thought.

"I was asked for my thoughts on the Resplandy paper as soon as it obtained media coverage.

"Most commentators appear to have been content to rely on what was said in the press release. However, being a scientist, I thought it appropriate to read the paper itself, and if possible look at its data, before forming a view...

"Because of the wide dissemination of the paper's results, it is extremely important that these errors are acknowledged by the authors without delay and then corrected.

"Of course, it is also very important that the media outlets that unquestioningly trumpeted the paper's findings now correct the record too. But perhaps that is too much to hope for."

https://www.nicholaslewis.org/wp-content/uploads/2018/11/A-major-problem-with-the-Resplandy-et-al.-ocean-heat-uptake-paper.pdf

REDUCING THE AMOUNT OF ACADEMIC GLOBAL WARMING FRAUD

Dear readers, it's OK to assume that if you see a UN or university spawned academic paper supporting Global Warming / Climate Change Caused By Human Behavior ... that it's a lie.

In this wonderful book you are reading we will cover the interesting subject of why this small group of a few hundred scientists keep lying.

For now we just give a little hint: money and political power.

Because money and political power are powerful motivators, we may pause to ask: Is it possible to stop, or at least to slow down the flood of fake Global Warming / Climate Change science?

A rapid and effective way to stop this mudslide of academic Global Warming fraud would be to cut off the federal funding. More than two billion dollars of federal money gets poured into the open mouths of opportunists who will gladly tell white lies for cash.

This is the result of the Obama regime working with the intent to utilize global warming as a power grab.

But while terminating funding is a great way to reduce fraud, because the American government is now largely a bureaucratic state, cutting off funding is easier said than done.

Departments are staffed with employees who can't be fired due to unions and federal regulations. Employees who are dedicated to global warming warriors or simply don't care about overspending. Or actually who don't care about anything. The checks keep rolling in every month no matter what.

Another approach to reducing the academic blood of false research papers would be judicial. Such papers cause material damage.

However, judicial methods are not simple and have a tendency to get into the "thought police" realm. A realm most people likely don't want to go.

But there is another way. And that is something which was taught to me by my Aunt Edith.

Starting at about five-years-old, she always brought me a bag of peanuts when she came to visit us. I think they were shelled, roasted, and salted if I remember correctly.

In any case, I absolutely loved them. I would run down the stairs at the first sign of her appearance.

But there was a catch. When I got the peanuts, I also got a loving pinch on the cheek.

Pinching children on the cheek is not done anymore and was a quaint American habit from her own childhood I suppose.

The problem was that because she loved me (and I, of course, loved her too), she squeezed sort of hard without realizing it. As in, when someone gives a relative a hug, the greater the loving intent involved, the warmer the hug.

The pinching hurt a bit, but I didn't say anything. Because I didn't want to say something mean to my aunt, and because the peanuts were part of the deal.

Eventually, I decided that I didn't want to get my cheek pinched any more, and I made a catastrophic decision. I told my mother about it.

That approach was very foolish, even for a five-year-old. I simply could have told Aunt Edith that it hurt a little, and asked if she could squeeze a little less.

And then offer my cheek with a loving smile. My five-year-old daughter today would have the smarts to do something like that right away.

But no... instead of taking the kind and diplomatic way, I told my mother.

That, in fact, stopped the pinching. It also stopped the peanuts.

While that was an unhappy event which I still remember today, I provide as a template for dealing with the flood of fraudulent global warming academic papers.

Stupidity has a price.

GLOBAL WARMING FRAUD ACADEMICS...AND APPLYING THE AUNT EDITH PRINCIPLE

Money, and to a much lesser extent, virtue signaling, is the motivation for writing global warming hysteria papers. It's mostly money.

But what if... along with the funding provided for the writing of these papers, there was a document. A small one-page statement, that the prospective authors would be required to sign...

> This tiny document states: If the paper contained fraudulent or manipulated data or if the conclusion contradicted the data in an obvious manner, the taxpayer money received for the paper would have to be returned in full. And that returned money would be donated to one of the following conservative organizations:
>
> **Americans United for Life, a large national women's pro-life organization**
>
> **The Trump 2020 presidential campaign**
>
> **The National Rifle Association (NRA)**

This approach is a serious one, and given that the Federal Government bloated regulatory agencies probably won't do anything, it is realistic. This is a real proposal which could have some good results.

And it is absolutely not "thought police" regulation.

It's not imprisoning anyone. It's not putting anyone before a firing squad.

It's simply requiring that, in exchange for government funding, there is a basic level of measurable accuracy in the papers that even a first year college student could produce.

So while the best way to reduce the endless little white lies in global warming support publications, is to cut off the government funding...meanwhile, **if the authors make a bad decision, meaning in this case if they lie, fudge data, misrepresent, cheat ... they give back the peanuts.**

CHAPTER 7

MORE ABOUT THE CRAZY SCIENTISTS OF THE GLOBAL WARMING FRAUD

THEY SAY THEY ARE THOUSANDS BUT THEY DON'T HAVE ENOUGH TO FILL A BUS

There is a tale spewed forth by Global Warming leaders, like a fountain of darkness... the wicked deception that their scientists number in the thousands.

That there is an army of academic darkness ready do do battle on behalf of Global Warming.

Here is the good news: there is no such army of darkness.

And here is how the deception happened.

Someone put the mathematical decimal point in the wrong place. The Global Warming / Climate Change "scientists" are always making mathematical mistakes, aren't they.

Let's estimate that there are about 200,000 scientists in the US who work in areas directly or indirectly related to climate. They are spread throughout the country. There are many more, but to help us understand the fraud, we'll just keep it at 100,000 for now. OK?

The leaders and the mob of Global Warming will want to give you the idea that out of the total, they have an army of scientists supporting them. About 97,000.

And of course, that's based on their fake University of Illinois poll.

But again, the Global Warming leaders and scientists have made a little teeny weeny math mistake... It's not 97% of them that believe in Climate Change and Global Warming caused by human activity, it's 97 scientists! (Less actually, because it's only 75, but let's be charitable and give them 97 as a number.)

Meaning for those of you on the Left, that out of the 100,000 scientists who work directly or indirectly with climate, 97 of them believe in Global Warming / Climate Change as something caused by human activity. Really.

And here is how we know it. It's not just a joke, it's real. And just to show you that this is true enjoy the following ride. Buckle your seat belts:

In 1998, Dr. Arthur Robinson, Director of the Oregon Institute of Science and Medicine, posted his first Global Warming skeptic petition, on the Institute's website (oism.org).

It quickly attracted the signatures of more than 17,000 Americans who held college degrees in science. Widely known as the Oregon Petition, it became a counter-weight for the "all scientists agree" mantra of the man-man Global Warming crowd....

Robinson's petition states a truth: "There is no convincing evidence that human release of CO_2, methane or other greenhouse gases is causing or will cause, in the foreseeable future, catastrophic heating of the Earth's atmosphere and disruption of the Earth's climate."....

(By the end of 2008, 32,000 scientists had signed the petition.) Now today it's double, triple or more.

And of course to be fair to all those little children of the Sunrise Foundation who are been brainwashed to believe that... "the scientists say...",

we should repeat that there are 97 scientists who actually do support Global Warming.

So the little mind-numbed robots would best be reprogrammed to say: "97 scientists say Global Warming is true."

But as we saw in the Introduction, for them, sadly, even that is not true. It's much, much less than 97 scientists. Much much less than even 75 scientists.

But let's say there are 97, to make the Sunrise children happy. And these 97 have, by the way, received financial support for their work. Piles of money. Very big piles of money. Seems strange, doesn't it? What does money payment have to do with Global Warming science? We are going to find out exactly in this exciting book.

LET'S TAKE A DETAILED LOOK, UP CLOSE AND PERSONAL, AT THE 97 GLOBAL WARMING FAKE SCIENTISTS

And why not start with the 10 heroes we covered in the last chapter. The "Computational" academic experts.

The article claimed to prove once and for all, we are heading into the bake ovens of Global Warming. And was said to be the knock-out to the sane and rational scientists they call "climate deniers".

And what a crazy term that is. Even for the political Left which can twist the language every-which-way. How can anyone be a "climate denier"? Crazy isn't it.

Anyway, as we already know, there turned out to be a sort of... what shall we say? A mistake? The math was all wrong, and there is no Global Warming.

Climate Scientist Nick Lewis pointed out the errors in Keeling's study in a blog post published Nov. 6, 2018 on climate scientist Judith Curry's website.

Lewis wrote that "Just a few hours of analysis and calculations ... was sufficient to uncover apparently serious (but surely inadvertent) errors in the underlying calculations."

Co-author of the now disclosed-as-"fake" paper, Ralph Keeling gave an "apology". He told a variety of media:

"When we were confronted with his insight it became immediately clear there was an issue there,"

"We're grateful to have it be pointed out quickly so that we could correct it quickly,"

"the combined effect of these two corrections to have a small impact on our calculations of overall heat uptake."

By saying this last half sentence, Keeling negates the whole so-called apology... by falsely putting the catastrophic math errors in terms of a "small impact".

"Unfortunately, we made mistakes here, I think the main lesson is that you work as fast as you can to fix mistakes when you find them."

That's a great lesson Keeling. How about this one: Do it honestly the first time.

But in the upside down world of Climate hysteria, Global Warming scientists are always right, no matter even if they are wrong. Because they are trying. This is apparently a common a mental disease of the political Left. To believe that what they actually do, doesn't matter. If they are "woke" and they try, then no matter what happens, even a major disaster, everything is OK.

So even though the study was a fake, all was forgiven. The Global Warming academic publication that was hyped by the Washington Post, New York Times, and the whole Democrat Media Complex as the final proof that Global Warming exists, turned out to be... as we said, well to be polite, scientific garbage.

It's fair to assume that all of the "Gang of Ten" actually read the paper they wrote and signed. And that they can do the basic math of a first year college student.

Paraphrasing the spokesman of the Gang of Ten:
"You see everyone, in science it's not important if you make a really really big mistake or not."

"What counts is if you issue a phony apology and say the important thing was not the catastrophic error you made in the paper and its totally false conclusions... Then everything is OK. And by the way, we're keeping the grant money.

That my friends encapsulates the Global Warming Scientists' approach to their work. It is political. It is financial. It's money and power. It's anything but scientific. And they keep the money, even if the paper is a total lie, and proven to be so.

The methodology of the 97 Global Warming scientists is this:

Get huge amounts of funding for research, get gigantic salaries, and then bend the facts to fit your fake narrative.

Global Warming fraudulent "scientists" world wide who have been caught faking math, providing false data, and presenting false conclusions to support Global Warming are only sorry about one thing. **That they got caught.**

And as we will see in this book. Lots and lots of the "Global Warming scientists" have gotten caught.

All righty, so in our countdown of fake scientists, the amazing academic paper of Dr. Laure Resplandy takes care of 10 of the 97 scientists, and there are now 87 left to take a look at.

TEN DOWN AND EIGHTY-SEVEN TO GO...

One of those remaining 87 is the dean of Environmental Sciences of the University of Michigan, Professor Johnathan Overpeck .

As California burned to the ground in the summer of 2018, he made the statement, that forest management was not involved, and Global Warming has caused the fires.

But let's look North, South, East and West. Let's look everywhere outside of California that summer. Everywhere else, where of course there was forest management... And we see... NO MASSIVE FOREST FIRES OUT OF CONTROL!

That's of course, something the Democrat Party bureaucrats in control can't quite understand. Other states and other countries use forest management so among other benefits, they can avoid big fires.

Later in this book we go into the California forest fires in detail, and how the Global Warming environmentalists stopped the clearing of dead brush and trees. Better to leave nature alone, they say. This crazy policy of radical environmentalism of the Democrat Party, has resulted in hell on earth.

Parents, if you have students who are studying environmental sciences at the University of Michigan, don't wait. Get on a plane, or in your car, or walk if necessary, and get your kids out of there! Why spend a fortune to fill your kids brains with Global Warming muck which will work against them when they try to get a job in a more sane, more rational, more accurate America?

And soon dear readers, after "Global Warming" is shown as bunk, how will these students get a job? In their resume and job applications will they write, "I was trained in fake science" ?

OK, now as we reduce the list of fair, accurate, smart, truthful scientists from the original 97, **that now leaves 86 "scientists"** who support Global Warming and who might be able to prove they are sane. Let's give them every chance. It's only fair.

There are many famous Global Warming scientists in Europe, what about them?

To find out, here is a fun-to-read article about the European Union academic Global Warming gangs, the first one at the prestigious London School of Economics. The London School of Economics is in fact a world leading university... so what could go wrong?

Here's what:
A top university in Britain has been caught stealing millions of taxpayer dollars from a federal budget that is chronically in the red just to produce phony global warming data.

A global warming research center at the London School of Economics received millions of dollars (pounds) from UK taxpayers after taking credit for research it did not perform.

The UK's Daily Mai noted further that the UK government provided £9 million ($11 million in US dollars) in funding to the Centre for Climate Change Economics and Policy (CCCEP) in exchange for research that was never done.

A number of papers that CCCEP claimed to have published to receive government funding were not even about global warming and were written before the organization was even founded, or written by researchers who were not affiliated with the center.

Worse, government officials never bothered to check CCCEP's alleged publication lists, saying instead they were "taken on trust," according to a recently released report.

"It is serious misconduct to claim credit for a paper you haven't supported, and it's fraud to use that in a bid to renew a grant," Professor Richard Tol, a climate economics expert from Sussex University whose research was reportedly stolen by CCCEP, told the Daily Mail.

"I've never come across anything like it before. It stinks."

The center's chairman since 2008 has been Nick Stern, a well-known global advocate for more policy action supposedly aimed at combating climate change.

In addition he is also the president of the British Academy, an invitation-only society that is reserved for the academic elite and which disburses grant money in the millions of pounds to researchers...

In recent days the CCCEP, which is jointly based at the London School of Economics and the University of Leeds, hosted a gala event that was attended by experts and officials from around the world.

The occasion: A celebration to mark the 10th anniversary of the Stern Review, a 700-page report detailing the alleged economic impact of climate change, a review that was commissioned by Tony Blair's government.

The massive paper claimed that the world must take immediate action to reduce greenhouse gas emissions or face dramatically higher costs in the future. The review exerted a great amount of power and influence on a series of British governments, as well as international organizations.

But now the report's contents and conclusions are in serious doubt.
https://dailycaller.com/2016/10/24/top-university-stole-millions-from-taxpayers-by-faking-global-warming-research/

and

https://www.naturalnews.com/055847_global_warming_phony_research_faked_data.html

OK, that little problem now leaves about 57 climate scientists who support Global Warming that might be sane or honest.

Let's look at some of the rest by reading this fun Daily Mail UK newspaper article about fraud, this time in America:

The Mail on Sunday today reveals astonishing evidence that the organization that is the world's leading source of climate data rushed to publish a landmark paper that exaggerated global warming and was timed to influence the historic Paris Agreement on climate change.

A high-level whistle blower has told this newspaper that America's National Oceanic and Atmospheric Administration (NOAA) breached its own rules on scientific integrity when it published the sensational but flawed report, aimed at making the maximum possible impact on world leaders including Barack Obama and David Cameron at the UN climate conference in Paris in 2015.

The report claimed that the 'pause' or 'slowdown' in global warming in the period since 1998 – revealed by UN scientists in 2013 – never existed, and that world temperatures had been rising faster than scientists expected.

Launched by NOAA with a public relations fanfare, it was splashed across the world's media, and cited repeatedly by politicians and policy makers.

But the whistle blower, Dr John Bates, a top NOAA scientist with an impeccable reputation, has shown The Mail on Sunday irrefutable evidence that the paper was based on misleading, 'unverified' data.

It was never subjected to NOAA's rigorous internal evaluation process – which Dr Bates devised.

His vehement objections to the publication of the faulty data were overridden by his NOAA superiors in what he describes as a 'blatant attempt to intensify the impact' of what became known as the Pausebuster paper.

His disclosures are likely to stiffen President Trump's determination to enact his pledges to reverse his predecessor's 'green' policies, and to withdraw from the Paris deal – so triggering an intense political row
.
https://www.dailymail.co.uk/sciencetech/article-4192182/World-leaders-duped-manipulated-global-warming-data.html

OK dear reader of this Primer On Global Warming that you hold in your hands... that now leaves about 37 "scientists" who support Global Warming who might possibly be honest or sane.

And you may have guessed it... thirty of those have been also wiped out for false data:

An independent audit of the key temperature dataset that is being used by climate models has exposed more than 70 problems with the data which render it "unfit for global studies."

Problems include zero degree temperatures in the Caribbean, 82 degree C temperatures in Colombia and ship-based recordings taken 100km inland.

The audit has revealed that "that climate models have been tuned to match incorrect data, which would render incorrect their predictions of future temperatures and estimates of the human influence of temperatures."

Furthermore, the Paris Climate Agreement adopted 1850-1899 averages as "indicative" of pre-industrial temperatures is "fatally flawed."

The entire Paris Climate Agreement has an agenda to eliminate effectively the advancement of society and attempt to reset the clock to the pre-Industrial Revolution.

This entire theory that before the Industrial Revolution, our planet's atmosphere was somehow pristine and uncontaminated by human-made pollutants has been also proven to be completely bogus.

Bubbles trapped in Greenland's ice has revealed that we began emitting greenhouse gases at least 2,000 years ago.

The Romans even constructed the first aqueduct was built in 312 BC because there was a serious problem with water pollution.

Seneca (c 4BC-65AD), the adviser to Nero, wrote in 61AD: "No sooner had I left behind the oppressive atmosphere of the city [Rome] and that reek of smoking cookers which pour out, along with clouds of ashes, all the poisonous fumes they've accumulated in their interiors whenever they're started up, than I noticed the change in my condition."

This new audit argues even the most simple basic quality checks had not been done on the HadCRUT4 data which is managed by the UK Met Office Hadley Centre and the Climate Research Unit at the University of East Anglia.

The audit exposed that estimates were made of the uncertainties arising from thermometer accuracy, homogenization, sampling grid boxes with a finite number of measurements available, large-scale biases such as urbanization and estimation of regional averages with non-complete global measurement coverage.

The audit has exposed the dishonesty in this entire scheme and it appears to be directed at the goal of reducing the population.

Anomalies it has identified include at St Kitts in the Caribbean, the average temperature for December 1981 was zero degrees, normally it's 26C. For three months in 1978, one place in Colombia reported an 82 degrees Celsius average – hotter than the hottest day on Earth. Then in Romania, one September the average temperature was reported as minus 46°C, which has never happened. The data showed that supposedly ships would report ocean temperatures from places up to 100km inland. The paper also points out that the most serious flaws identified was the shortage of data.

For the first two years, from 1850 onward, the only land-based reporting station in the Southern Hemisphere was in Indonesia. Then there were ship observations at the time but Australian records had not started until 1855 in Melbourne, behind Auckland which started in 1853. This data appears to have been just made up.

https://www.zerohedge.com/news/2018-10-12/armstrong-independent-audit-exposes-fraud-global-warming-data

OK dear readers... THAT LEAVES seven "SCIENTISTS" IN THIS WORLD OF OURS, WHO SUPPORT GLOBAL WARMING that might be honest or sane.

And the final seven are wiped out as we will now read in these two articles exposing their fraud:

"Global warming alarmists are scrambling to save face after hackers stole hundreds of incriminating e-mails from a British university and published them on the Internet.

The messages were pirated from the Climatic Research Unit (CRU) of the University of East Anglia (**UEA**) and reveal correspondence between British and American researchers engaged in fraudulent reporting of data to favor their own climate change agenda.

UEA officials confirmed one of their servers was hacked, and several of the scientists involved admitted the authenticity of the messages, according to the New York Times.

The article opined, "The evidence pointing to a growing human contribution to global warming is so widely accepted that the hacked material is unlikely to erode the overall argument."

Climatologist Patrick J. Michaels challenged that position. "This is not a smoking gun, this is a mushroom cloud."

The e-mails implicate scores of researchers, most of whom are associated with the UN's Intergovernmental Panel on Climate Change (IPCC), an organization many skeptics believe was created exclusively to provide evidence of anthropogenic global warming (AGW).

Among the IPCC elite embarrassingly, if not criminally, compromised is Phillip D. Jones, a Ph.D. climatologist at the University of East Anglia whose work figured prominently in the IPCC Third Assessment Report of 2001. Jones also contributed significantly to the IPCC Fourth Assessment Report in 2007 (AR4), but he failed to follow through when skeptical investigators asked to review raw data associated with that report.

They announced intent to use UK Freedom of Information laws to obtain the data, so Jones sent the following e-mail to one of his collaborators: "Mike, Can you delete any e-mails you may have had with Keith re AR4? Keith will do likewise.... Can you also e-mail Gene and get him to do the same?... Will be getting Caspar to do likewise."

The Mike in this message is Michael Mann, professor of meteorology at Pennsylvania State University, whose influential "hockey stick" graph warning of pending global warming eco-catastrophe was found by a congressional investigation to be fraudulent.

Mann seems to be the Walking Dead. He keeps on coming. No matter how many times his fraud is exposed he just keeps hopping to the next one. The most recent effort was his input into the "Our Planet" Netflix series, which among other things managed to kill innocent walruses by using drones which drove them off the edge of a cliff.

The series of course blamed "Global Warming" for the walrus "suicides". But it was panic caused by the video drones of the series which led to the catastrophe.

Now back to Michael E Mann, in this previous scam... In another correspondence about AR4 labeled HIGHLY CONFIDENTIAL, Jones contacted Mann regarding research critical of their global warming platform. "I can't see either of these papers being in the next IPCC report," wrote Jones.

"Kevin and I will keep them out somehow — even if we have to redefine what the peer-review literature is!"

Mann received another incriminating e-mail from Dr. Kevin Trenberth, a New Zealander now with the University of Colorado and Head of the Climate Analysis Section at the National Center for Atmospheric Research. "The fact is we can't account for the lack of warming at the moment and it is a travesty that we can't." An incredulous Trenberth simply blamed "our [inadequate] observing system."

Yet he and his colleagues are now dodging the "Climategate" bullet, indignant that global warming skeptics are supposedly taking their comments out of context. One wonders if they might be referring to a message from Jones who wrote about a statistical "trick" he used to "hide" data.

Or perhaps they mean Mann's reference to climate change skeptics as "idiots."

https://www.thenewamerican.com/tech/environment/item/6748-ipcc-researchers-admit-global-warming-fraud

And finally dear reader, the number of sane or honest "scientists" who support Global Warming is taken to zero. Read this from a Daily Caller and Natural News articles:

"A rational review of global warming data as unveiled systematic scientific fraud to alter temperature data in support of the global warming false narrative. This is the largest discovery of scientific fraud in the history of science, and it shows that "global warming" and "climate change" are elaborate science hoaxes rooted in fraud, not fact.

As The Daily Caller and Natural News reports:

A new study found adjustments made to global surface temperature readings by scientists in recent years "are totally inconsistent with published and credible U.S. and other temperature data."

The purpose of the widespread fraud has been to achieve "consensus" by exposing scientists to fake data that appear to show a catastrophic rise in average global temperatures. It's all being done to support the moneymaking scam of carbon taxes that enrich fraudulent science hoaxers ike Al Gore who are raking in billions of dollars from carbon tax schemes and oppressive government regulation of carbon emissions.

The fraudulent warming data are then used as a basis for climate modeling software systems that extrapolate the fraudulent data to predict "climate doomsday" for the planet.

This is where delusional scientists like Stephen Hawking lose their minds and claim that Earth will be transformed into Venus with temperatures over 800 degrees (F)

and sulfuric acid rain. In truth, humanity couldn't achieve such large-scale terraforming outcomes even if we tried.

"Nearly all of the warming they are now showing are in the adjustments," Meteorologist Joe D'Aleo, a study co-author, told The Daily Caller News Foundation in an interview. "Each dataset pushed down the 1940s warming and pushed up the current warming."

The greatest science HOAX in human history

The entire climate change / global warming narrative is an elaborate science hoax that's being continually faked through the ongoing, systematic alteration of temperature data in order to "conform" with the false narrative.

At every level, it is Orwellian science at its worst, complete with its own Ministry of Truth that pushes its false narrative through the fake news media. Many scientists are deliberately participating in the fraud, knowingly working to alter temperature data in order to prop up their delusional narratives that collapse under the slightest scrutiny.

This is why all rational skepticism about climate change is shamed and silenced: Because the fake science narrative cannot withstand scientific scrutiny.

Thus, its proponents declare "the science is settled," meaning no discussion or dissent shall be allowed (because if it were, the fraud would be quickly exposed).

This is how "science" got hijacked by climate change cultists who share more in common with dogmatic, tyrannical CULTS than anything that could be honestly labeled "science."

https://www.naturalnews.com/2017-07-25-global-warming-bombshell-systematic-science-fraud-revealed-in-alteration-of-temperature-data.html

What kind of words could describe the Global Warming / Climate change "scientists" who lie, cheat, and steal? Who have come to steal, kill and destroy?

We can't even imagine, but probably words so nasty they could not be put in this book. Soon these fake "scientists" will be relegated to the dumpster of academic history, known as "THE HALLOWED HALL DUMPSTER OF SHAME".

But perhaps there is one, one Global Warming / Climate Change From Human Activity "scientist", that is not insane or a scientific criminal groveling for money or political power ?

Well, dear readers, perhaps somehow somewhere, we can find a few scientists who support Global Warming, and who are sane and honest? Can we find even a few? Let's try our best. One more try.

Let no one say, this book has been unfair. Onward in our search for an accurate and honest and existing Global Warming proponent scientist.

What about those Global Warming scientists at the best universities in the world? Let's take a look at for example, Harvard and Yale.

OK, what about the Global Warming / Climate Change hysterics there?

Well... amazingly enough, in November 2018, a group of "scientists" from Harvard and Yale, suggested a plan to deal with their fears of a Global Warming disaster:

Spray a chemical called sulfates into the atmosphere. The sulfates, proclaim the geniuses, will absorb heat from sunlight and the planet would be cooled.

Brilliant? Right?

Supposedly brilliant universities, supposedly brilliant scientists... perhaps finally in our search we have found some Global Warming / Climate Change scientists who are not insane or con men and women writing lies for the money.

Let's find out in this final search effort to find some that are sane and not liars with an agenda.

We are now focusing on what are supposedly the most brilliant universities, and on the supposedly brilliant scientists, and upon their supposedly brilliant ideas...

so what could possibly go wrong with spraying sulfates into the atmosphere?

First of all, isn't Global Warming itself supposedly caused by mankind (which for the sake of political correctness we will call "humankind") pumping pollutants and gunk into the atmosphere?

And now they want to pump a huge amount more into it?

Secondly, sulfate has a teeny weeny tiny little environmental side effect.

It effects the water of our world. In short, it produces ACID RAIN !!!

Acid rain is sort of similar to our natural rain, except that chemical changes have occurred from pollutants in the atmosphere that make the rain not friendly, but an enemy.

Instead of beautiful, nourishing, cleansing water, the Harvard-Yale water would be something a bit different.

When exposed to the sulfates from Harvard and Yale, we get... an acid bath ! An acid drink. Acid to drink for both us, the animals that the environmental hysterical Left claims to love so much.

And acid to drink for all the plants, for the forests, that the same tyrannical hate based environmentalists of the Left claim to love so much.

Yes, buckets of acid pouring down from the skies, courtesy of Harvard and Yale Global Warming / Climate Change hysteric "scientists".

Acid rain damages plants, animals, and surfaces, such as car paint, buildings, and even our very own surfaces: hair, skin, eyes and little things like that.

Acid rain also ruins the water supply of the earth. We'd be drinking acid. When we wash our little children's faces they could go blind.

Oh Harvard and Yale, how you have fallen.

Parents, if you have kids studying at Harvard or Yale, get on the next plane, car, bus to those places and drag your kids out of there, to save their minds from brainwashing and their skin from acid rain.

What a crazy idea from the best of the best Global Warming scientists.

Now in addition to harming children who end up studying with them, the Global Warming / Climate Change activist "scientists" seem focused on destroying the trees of our planet.

Because Global Warming hysterics claim to love trees, it's useful to point out that acid raid kills trees. Yes that wonderful acid raid courtesy of Harvard and Yale "scientists" would kill the trees.

Here is a photo of what was a beautiful forest that was subject to acid rain, the same type that would be caused by the Harvard and Yale geniuses.

Here in this disturbing photo are the results of acid rain on trees. And is exactly what the geniuses from Harvard and Yale would cause by their unhinged belief in Global Warming and their apparent total lack of understanding basic climatology and even high school chemistry:

Spooky, isn't it. If you know someone who is thinking of attending Harvard or Yale in the sciences, please warn them. Attending might produce liberal brain toxicity.

And oh my goodness, there is more.

In addition to wiping out the trees, sulfates sprayed into our atmosphere would cause a change in clouds. The size of the water droplets would change, bringing more rain, and the dreaded flooding that Global Warming hysterics are yelling about in the first place.

Cloud height would change and bring results we can't even imagine.

So, why not cross these Global Warming scientists from Harvard and Yale off the list of sane and honest people too. Even CNN wrote that this idea of spraying sulfates into the atmosphere "sounds crazy". And if CNN says it's crazy, it is.

Well done Harvard and Yale climate change scientists! Doing your best to destroy the planet by making mistakes that even a high school chemistry student would not make. The science of the Left is a mental illness.

AND THEN THERE WERE NONE

If dear reader that somehow left a few scattered about just a few dishonest "scientists" who support Global Warming fraud... they disappeared when sadly they froze to death during outdoor "save the planet" protests in 2018.

Yes, in the 15 to 20 degree weather and blizzards which engulfed the United States around Thanksgiving 2018, the few remaining Global Warming fraud scientists were out protesting in their shorts and flip flops In the tradition of Al Frostbite Gore, they said ...Good bye.

This amazing unedited (really) photo shows the Witch of Freezing, coming to get those crazy Global Warming & Climate Change so-called "scientists". And the Witch succeeded big time.

Happy Thanksgiving Climate Change Hoax-loving scientists. Now extinct.

And then there were none. So dear readers... Thus we can say with confidence, that no scientist exists anywhere that supports Global Warming / Climate Change that is not either frozen or recovering from a Leftist lobotomy, or crawling after money for research horror stories.

Meanwhile, there are hundreds of thousands of real scientists who find the Global Warming / Climate Change Caused by Human Behavior fraud an affront. They are fighting the battle for truth. The Democrat Media Complex works to keep them hidden, to censor them. And that we cover later in this exciting book.

THE FOUNDATION OF GLOBAL WARMING IS LIES... FOR LEFTIST READERS, IT'S TIME TO GET ON THE TRUTH TRAIN

As we have seen, and will see more, in this guidebook to sanity and truth in climate, fake Global Warming data, rampant fraud and lies are the underpinning of the Global Warming / Climate Change movement.

So dear reader, here is a chance to get on the truth train.

We are heading into an ice age, and this wonderful little book you have in your hands will provide the information why it is starting and what is going to happen to our society and country when it occurs.

We have had them before on earth and as recently as from 1300 to 1800, so we know what happens.

And we will see also how the Global Warming / Climate Change From Human Behavior hysteria is keeping the country from getting ready from the coming freezing catastrophe. How the leaders from Al Gore to AOC to Zach Exley and Bekey Bond, and the puppet master billionaires behind them, are doing their best to drive the nation into national suicide.

Because at a time we should be devoting national and personal resources to get ready for the coming Ice Age, we're preparing for a non-existent Global Warming crisis. It's going to be very very cold, so why not get one of these wonderful ICE AGE sweatshirts:

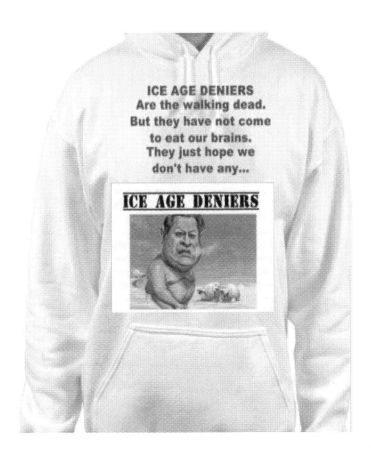

The year will be 2025... it's going to get a bit chilly... get ready.

CHAPTER 8

GLOBAL WARMING, AND GLOBAL COOLING SUCH AS ICE AGES, DO OCCUR.

BUT ARE THEY CAUSED BY HUMAN ACTIVITY?

Well, we already know, that the amount of CO_2 in the air has no effect on heating or cooling the earth.

But let's for the sake of our friends on the Left, that it does, let's pretend CO_2 does have a warming effect (which it doesn't, but let's pretend).

Even in the case that CO_2 could cause warming, there is some more bad news for those who believe that Climate Change / Global Warming is caused by human activity:

Of all the CO_2 released into the atmosphere every year, less than 3.4% is caused by humans.

Putting that another way: of all the CO_2 released into the atmosphere every year 98% comes from other sources than human activity.

The rest is caused either by plants and other living beings that "exhale" CO_2, or by decomposition of organic compounds.

Now because this book is partially intended for those on the political Left, and Left wing college students, I am sorry to have used such big words as "organic compounds" or "decomposition". (Just kidding)

But what it refers to are things like leaves from trees. After they fall from trees in the Autumn, they decay. "Decay" is a fancy word for "rot". And when things which were once alive decay, they release guess what... CO_2.

When we throw out unused food into the garbage, it decays. When a tree falls down in the forest, it decays. Anything that dies, decays. And that decay process releases CO_2 into the atmosphere. A lot of it.

And when living things are alive, they exhale CO_2.

So this is really bad news for those who are part of the Climate Change / Global Warming false religion. When they are alive.. plants, trees, and people and animals "exhale" CO_2.

And when they die, they begin decomposition, an even longer word, which also releases CO_2.

So out of all the new CO_2 in the world every year, only about 3.4% is caused by human activity, such as travel, manufacturing things, heating homes, cooking... the Leftist billionaires traveling around on their private jets and SUVs, breathing, soft drinks and so on.

So even if all human activity would suddenly stop, it wouldn't make a bit of significant difference to the CO_2 level of the earth.

Now dear reader, we can stop pretending that CO_2 has any effect on temperature of the earth. It was fun and useful to pretend, so we could understand better that human activity has about zero effect on CO_2 levels on the planet. But the effect of CO_2 on temperature is zero.

And a tiny little footnote. The earth has been around for billions of years. It has had many periods of extreme cold and extreme heat. Massive ice ages, massive very hot periods. Long before human beings were every around. All this results from the sun. If the sun in a hot phase or a cool phase. That's it in just a few words.

We're now heading into a cool phase for the sun. We will read the scientific evidence in this book. And that cool solar phase means one thing: ICE AGE.

We've picked that date of 2025 for a reason. A best estimate. But if not then, 2035 for sure. Or next year. But we're on our way over the climate cliff into the Ice Age valley of cold temperatures.

CHAPTER 9

WHAT IS A "GREENHOUSE GAS" ?

Everybody uses the term, "greenhouse gas", but hardly anyone who is in the Global Warming & Climate Change false religion knows what it actually is.

Now we are going to tell them.

For those of you who spend a lot of time protesting and yelling and who are filled with rage about all things environmental and political, maybe you have not had time to visit an actual greenhouse.

Carrying all those protest signs, sitting on the floor in sit-ins, yelling while walking down the streets is very time consuming. So maybe you were never inside an real greenhouse.

Therefore for those readers on the political Left, the first step is to explain what a greenhouse is.

People discovered that even when it's cool or cold outdoors, they can still grow plants. They simply need to put the plants inside a house (which can be large or small) made of glass walls and ceilings.

The sun shines on that glass house and heats up the inside of the glass house.

Some greenhouses are small and used by gardeners, but some are really really big, and used for growing food commercially. They can be as big as a football field or even bigger. They grow hydroponic vegetables and plants, and we will discuss that later. Dear reader, I've had many years experience in hydroponics and designed large systems. Greenhouses are good. Not bad.

I know because I have designed and built some big ones. Hydroponic food growing centers for lettuce and tomatoes.

Sometimes greenhouses are all glass, and sometimes they have concrete, wood, or brick walls and a glass roof. They are usually used in cold places.

OK, now that we know what greenhouses are, we can find out what a "greenhouse gas" actually is.

Just like the glass on a greenhouse can catch the heat of the sun and hold it in the greenhouse, gasses can do the same thing. The sun shines through the glass and heats the greenhouse building. It also heats the glass itself, which radiates more heat into the greenhouse.

A lot of the heat stays inside it as long as the sun is shining on it. And greenhouse gasses do the same thing for our planet.

We are very very lucky.

We would be dead without them.

Little pieces of unhappy ice.

These wonderful and vital natural gasses are found throughout our atmosphere. And there are many of them. The main greenhouse gas is not CO2, but is … water vapor. So if we really want to eliminate greenhouse gas, then the target should be clouds.

But surprise, surprise, natural greenhouse are good, not bad.

The main greenhouse gasses found in our air are:

- water vapor (by far the most)
- carbon dioxide
- methane
- nitrous oxide
- ozone

There are also many many other greenhouse gasses, and we have a full list of them coming up.

To impress any of your comrade Global Warming protesters, if you are one of them, just memorize the whole list and recite it while you are on a sit-in.

Anyway, just as a piece of glass in a greenhouse catches and holds heat from the sun inside the greenhouse, the "greenhouse gas" does the same thing. Except instead of holding the heat in a greenhouse, it holds the heat for us on earth.

These gasses transmit heat from the sun to the earth. And then they keep the heat trapped to some extent in our atmosphere. Just like a glass greenhouse absorbs heat. Wow.

According to people who believe in Global Warming & Climate Change by human actions believers, **greenhouse gasses and what they do is really really scary.** A sort of greenhouse glass roof in the sky made of gas that heats the earth.

And which, they say, is going to bake us like little cookies.

But as usual, what seems bad news to climate hysterics is actually good news.

Why?

If we didn't have greenhouse gasses surrounding the earth, much of the heat which we need to live, would escape out into space. Al Gore's nickname "Frostbite Al" would come to reality sooner than expected.

Without greenhouse gasses to protect our planet and our lives, the average temperature on the earth's surface would be 0 degrees F (-18 degrees C) which means we would all die within months for lack of food, water, warm shelters and so on.

Think about it... When it's always 0 degrees in America, all year long, no plants will grow outside except for some very rugged evergreen trees and lichens (those funny little green things that grow on forest rocks and tree bark) that we can't eat.

That means no more food for us or the animals without greenhouse gasses because of the low temperatures.

So, that "scary" greenhouse in the sky is actually good for you, and saving your life ! Without greenhouse gasses we would freeze and quickly.

It's not really scary at all. It's wonderful.

Just think about it. No greenhouse gas? Life on earth ends as we know it in freezing hell.

Because the belief in Global Warming / Human Caused Climate change is actually a religion (not a science), why don't the Global Warming hysterics pray every night, thanking whatever they pray to, for greenhouse gasses.

For keeping us, the animals, and plants alive on this earth. We couldn't get along without the greenhouse gasses.

CHAPTER 10

THE REALLY STRANGE THING ABOUT THE GREENHOUSE GASES COMPARED TO OTHER GASSES

AND THE AMAZING TRUTH HOW LITTLE CO2 IS AROUND

There are many gasses that supposedly have a greenhouse effect and here they are. Some are natural and some are produced by chemical processes. It is quite odd that the fake Climate hysterics chose CO2 as their target. A harmless, in fact beneficial, natural molecule. Oh well, who every said the the Climate hysterics were rational.

Get ready for a long list, but you paid for this book, so you should get many pages of paper. When the coming ice age hits, you can then burn these pages to keep warm. What a great book! Dual purpose!

These greenhouse gasses are in addition to those listed before in Chapter 6 (remember that the main one is water vapor, some of which we see in clouds):

GREENHOUSE	GASSES	
Methane	12	25
Nitrous Oxide	114	298
CFC-11	45	4750
CFC-12	100	10900
CFC-13	640	14400
CFC-113	85	6130
CFC-114	300	10000
CFC-115	1700	7370
Halon-1301	65	7140
Halon-1211	16	1890
Halon-2402	20	1640
Carbon tetrachloride	26	1400

Methyl bromide	0.7	5
Methyl chloroform	5	146
HCFC-21	1.7	151
HCFC-22	12	1810
HCFC-123	1.3	77
HCFC-124	5.8	609
HCFC-141b	9.3	725
HCFC-142b	17.9	2310
HCFC-225ca	1.9	122
HCFC-225cb	5.8	595
HFC-23	270	14800
HFC-32	4.9	675
HFC-41	3.7	92
HFC-125	29	3500
HFC-134	10.6	1100
HFC-134a	14	1430
HFC-143	3.8	4470
HFC-143a	52	4470
HFC-152	0.6	53
HFC-152a	1.4	124
HFC-161	5.5	12
HFC-227ea	34.2	3220
HFC-236cb	13.6	1340
HFC-236ea	10.7	1370
HFC-236fa	240	9810
HFC-245ca	7.6	1030
HFC-365mfc	8.6	794

HFC-43-10mee	15.9	1640
PFC-14	50000	7390
PFC-116	10000	12200
PFC-218	2600	8830
PFC-318	3200	10300
PFC-3-1-10	2600	8860
PFC-4-1-12	4100	9160
PFC-5-1-14	3200	9300
Sulfur hexafluoride	3200	22800
Nitrogen trifluoride	740	17200

*100-Year GWP relative to carbon dioxide (CO2)

Once those of you who are on the political Left have these memorized, as well as water vapor, carbon dioxide, methane, nitrous oxide, and ozone, you will be ready for street protests or sit-ins. So get busy memorizing.

As we can see, there are a whole lot of greenhouse gasses. Not just CO2, not just carbon dioxide.

So it's sort of silly, isn't it to focus on just one of the greenhouse gasses, if they cause overheating of the earth?

Particularly since CO2 is less than 3.62% of the total greenhouse gasses in the atmosphere? And since people don't cause hardly any of that CO2...

it means the whole Global Warming /Climate Change movement is sort of, well, crazy.

It's sort of like living in New York City, and making a law against sharks being able to roam freely in the city swimming pools. It's picking a strange target for the law, because there are very very few sharks there. Choosing CO2 as the evil gas and the target is just like that.

If the "movement" wants to get rid of a greenhouse gas, the most abundant by far is.... water vapor.

Yes, all that past and coming action from the Democrat Party, Justice Democrats / Democratic Socialists and its followers should really be focusing on eliminating clouds.

And other water vapor in the air. Because that accounts for 95% or more of the global warming gasses.

Just think of how much good we could do by eliminating all water !

So get in line behind Al Gore, behind the The Science Jerk, and AOC, and go it it!

Try to get the water cloud footprint down to zero if you can. Goodbye pretty clouds.

These must be disturbing things for readers on the Left, and important news for conservative constitutional Republicans who oppose the fraud.

Why are we spending all this time and trouble in this booklet to see what greenhouse gasses are?

Well, first, it's nice to know when take a position on something, that we know something about it.

Even for the Global Warming hysterics. People out protesting against greenhouse gasses need to know the whole list of them to do it right. Remember, no more protests until you memorize the whole list. OK?

And after those of you who are still in the false religion, have memorized all the greenhouse gasses, then take a step up.

Why not have really really big protest signs with all the greenhouse gasses written on them? And with clouds as the #1 enemy which causes Global Warming fire, death, and destruction, if greenhouse gasses do that.

And for everyone else who reads this book, who are living in reality, there is a "feel good" about knowing what greenhouse gasses actually do to keep life on earth, and what the main ones are. But there's another reason it's nice to know what greenhouse gasses are. And here is that reason explained in the next chapter.

CHAPTER 11

SOME LITTLE NUMBERS TELL A BIG TRUTH

I promised those readers on the political Left, we would not use long words or big numbers, and I've mostly kept that promise.

Here we need to look at some numbers, but they are small numbers so no problem.

If by now you don't get it that Global Warming / Climate Change From Human Activity is a scientific fraud, these little numbers will turn on the light bulb for you.

Although the list of greenhouse gasses is a long one, the total amount of greenhouse gasses in the atmosphere is less than 3% of the total gasses of the atmosphere.

Greenhouse gasses are a drop in the bucket of the total air.

And CO_2 is a drop in the bucket of greenhouse gasses.

For those on the Left, for the majority of Leftist college and university students, so you don't get a headache thinking about his, it's written here in an easy to remember way.

Tell everyone you know. And by knowing the truth, former Global Warming fraud supporters can go from zero to hero.

Here is the nice summary to help:

Of the entire atmosphere the total
GREENHOUSE GASES are just 2%

CO_2 is only 3.62% of the greenhouse gasses.
So of greenhouse gasses in the atmosphere, a drop in the atmospheric bucket, just 3.62% of that little drop in the bucket is CO_2

And now here the most fun of all:

Of every bit of CO2 that is in the atmosphere, **only 3.4% is produced by human activity.**

Yikes! People make hardly any CO2 !

Think about it, those attending Leftist colleges and universities. Come on, you can do it.

Here is a chart to help us:

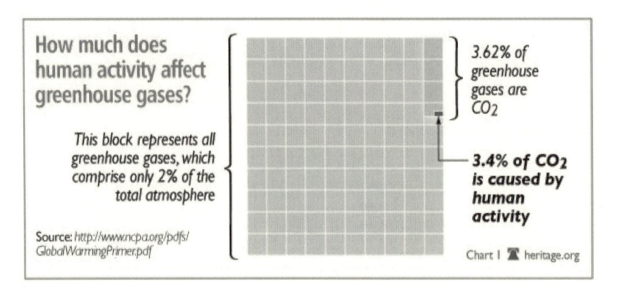

It means that in the atmosphere, of the total gasses:

- Greenhouse gasses represent **2.0%** of the total atmosphere. Not much is it?

- *Of the total gasses in the atmosphere,* CO2 itself is just **0.07%**

- *Of the total gasses in the atmosphere, CO2 caused by human activity* is only a tiny **0.002%** **Yikes, it's almost nothing.**

Ding, Dong, ring the bells of victory. The Global Warming battleship of lies is now sunk. And by these tiny little numbers.

As we see, the amount of CO2 in the world caused by human activity compared to the entire atmosphere is tiny. Even if the cow farts of AOC are added in, it doesn't change anything.

Why then did we bring up all those numbers, particularly when we decided to use as few long numbers as possible, so not to intimidate those on the Left who are part of the Global Warming false religion? And their "scientists" who appear to have severe and unending math problems.

We brought up the numbers just to show, once and for all, that all greenhouse gasses together are a teeny weeny part of the total gasses in the atmosphere of the earth.

And that CO2 is a teeny weeny part of the greenhouse gasses in the world.

And even worse for those who believe Global Warming / Climate Change is caused by human activity producing CO2, is almost zero.

CO2 itself, and particularly CO2 produced by human activity, is just a teeny weeny, teeny weeny, teeny weeny part of the atmosphere.

Remember, of all the CO2 on earth, just 3.4% of that is the result of our factories, cars, jets, fossil fuels and so on.

Poor little fossil fuels. They really aren't making much CO2 at all. And they get blamed for so much, unjustly.

Most of you who are reading this already knew and knew that Global Warming hysteria is a fraud. But some of were on the political Left and didn't know that.

So for those on the political Left, let's suppose you've had a hard day protesting against Global Warming. Maybe you did it for free. Maybe your expenses were paid by a group financed by George Soros or Tom Steyer or one of the other Leftist billionaires. Maybe Jeff Bezos or Michael Bloomberg. Funders of the new professional protesting class of people.

Anyway you've been yelling, carrying around heavy signs, and walking. And to relax after your hard day protesting while the rest of us were at work, you take a field trip to the ocean shore. You're in California of course.

And while you are there, you drop your Starbucks latte into the ocean by mistake. That liquid, the cup of Starbucks latte in the ocean, is about the same amount as CO_2 produced by human beings is in the atmosphere. The latte is the CO_2 and the ocean is the atmosphere. It's not much, is it. Wow !

So all that Global Warming / Climate Change protesting going on, is about... well... nothing.

CHAPTER 12

BUT WHAT ABOUT THE GREEN NEW DEAL ?

WON'T ALTERNATIVE ENERGY LIKE WINDMILLS AND SOLAR PANELS HELP US ?

MAYBE THAT'S SOMETHING GOOD WHICH IS IN THE GLOBAL WARMING MOVEMENT ?

Sorry to disappoint the political Left. But the alternative energy equipment in which they put their hope... will not work in either very warm or very cold weather. So if Global Warming is real, their believed equipment will not be functional. It will be expensive piles of high tech trash.

Here is the story:

Why Green Energy Won't Work In The Alarmists' Future

My background is as an engineer working in alternative energy for over 25 years.

Alternative energy projects I have done and equipment I've designed are in the US, Canada, Malaysia, South Korea, China, Russia, Israel, and the EU.

The specific production-type technology is hydropower, but also wind, solar photovoltaic and solar thermal. And for some strange reason, I'm a specialist in the effects of unusually cold ambient temperatures on infrastructure. And cold is the direction we are headed, so perhaps this might be a useful area.

When I look at the Green New Deal, and its reliance on what they call "alternative energy," I don't know whether to laugh or cry.

If the outlandish tale of global warming was to happen, these so-called alternative energy systems simply don't work in high-temperature environments.

Let's go briefly through their two major hopes for alternative power: photovoltaic and wind.

Solar electric panels don't work well in hot weather. That's why we don't see any in places like the United Arab Emirates / Dubai.

Solar panels are from the start rather inefficient, at best 22% or 23%.

Bad news for AOC: As the temperature rises, they become less and less efficient. When the temperature **of the panel** reaches 42 degrees C (107.5 degrees F), they begin losing the little efficiency they have and fast.

In places like Arizona, where people make the effort to USE them, they need to be cooled by running cold water in pipes underneath them to make them function reasonably well.

But here's the kicker in determining the efficiency of photovoltaic panels at high temperatures. **The temperature of a photovoltaic solar panel is usually about 20 degrees C (36 degrees F) hotter than the ambient air.**

This means that the loss of efficiency actually begins when the air temperature is 22 degrees C (71.6 degrees F).

For every degree above ambient air temperature of 22 degrees C, the panels will lose 1.1% of their peak electric output.

Just think about the hysterical projections of the Al Gore – AOC followers, and we see that in their nightmare world, the panels will be operating at efficiency levels approaching zero.

Just when it seems that things can't get worse for the global warming movement, it does. Not only do the PV panels drop in efficiency as the temperature increases... their product life decreases.

Right now, with the most advanced PV technology (China of course, which is another problem), the best solar electric panels have a lifespan of about 25 years **absolute maximum**.

After that, the only place for them is the garbage dump.

They are not recyclable at all. Except for the little aluminum frames around each one.

And after 25 years of use... each solar electric panel has to be thrown into the garbage dump...

Thus, we will end up in the future with new mountain ranges in America. Made of burned-out photovoltaic panels. Every 25 years, every single one of them is going to become part of the new garbage mountain range.

If the temperatures projected by the global warming hysterics would actually occur, the product lifespan of the panels would be reduced to a few years. And those years, the panels would be operating at low efficiency.

Yet, it gets even worse for AOC and Bernie Sanders' PV panels. The electronic control equipment which controls and directs the harvested electricity conks out at about 40 to 45 degrees C (104 to 113degF). Which is a cool day in the doomsters' global warming future.

While these environmental temperatures may seem too high to happen, the equipment itself, just like solar panels, tends to become even hotter than ambient air.

So much for the solar electric panels. Goodbye.

What about the wind turbines?

The turbines themselves and electronic equipment which controls electrical production and distribution matters are housed at the top of the single column towers, in egg-shaped shells made of metal.

They are called nacelles. In the hot global warming future, these egg-shaped (or shoebox-shaped with rounded corners) metal containers **are going to become bake ovens for the equipment they contain.**

First to go is, of course, the electronic control equipment. As mentioned, that conks out at about 40 to 45 degrees C.

Next to go is the turbine itself. Metals will slightly change shape and start to exceed their required tolerances, and the lubricants won't function. But it gets even worse for the global warming hysterics…

Along with hot temperatures in the alarmists' nightmarish future, hysterics are promising wind storms, tornadoes, and hurricanes.

These will rip the turbine blades to shreds. They do not have the ability to withstand the force of such high winds.

A fiberglass or carbon/glass composite blade is often about 50 meters in length, weighs about 4,000 kg and will snap like twigs in the high winds expected by the global warming activists.

And as a footnote, hydropower would have a less difficult challenge in the hot temperatures. However, the electronic

control and distribution equipment needed for them to work will conk out at about 40 to 45 degrees C.

The Ship Of Truth

The ship of truth floating in the sea of debate is being kept afloat by tens of thousands and hundreds of thousands of scientists who recognize man-made global warming as a scientific fraud. And by many groups and organizations and websites also fighting the fight.

If AOC and friends are hoping for alternative-energy equipment to help solve their delusional problem, they will be very disappointed.

They and the so-called global warming "scientists" might as well roam cow-filled fields and pastures with empty plastic water bottles and corks, hoping to capture cow methane as a fuel for their future.

And in fact, they are going to need that methane to heat their homes. Because about 2025 we are over the cliff into a new Ice Age.

CHAPTER 13

THE NEXT ICE AGE.
COMING TO YOUR NEIGHBORHOOD VERY SOON..... DON'T MISS IT!

Since we will soon be skiing down the slope of an Ice Age (at least a Little Ice Age), it might be interesting to know what causes them.

No, it has nothing to do with CO2 or human activity...

Ice Ages are caused by the sun. Or better to say, by the lack of sun.

The temperature on the earth is determined by the sun. And the sun itself becomes hotter or cooler in long cycles.

When the sun is in one of its warmer cycles, the surface of the earth is hot.

When the sun is in one of its cooler cycles, the surface of the earth is cooler and can enter an ice age.

To make it clear as day, think of it this way:

If the sun suddenly goes out, if it suddenly stops producing any heat, the earth would be a block of ice within hours. And we would all be little blocks of ice with it.

On the other hand, if the sun suddenly powered up with enormous energy, the earth would be burned to a crisp. And we would all be like little toasted marshmallows.

Since 1800 the sun is in the middle between freezing us and burning us up. But the sun doesn't just stay right in the middle. It's a star, and it has hotter periods and cooler periods.

When the sun enters a cooler period, called a solar minimum, when it enters into a cool cycle, then here on earth we have an Ice Age.

It's nice to understand that. And we can be happy that we know it.

But the bad news is that the sun is entering a much cooler cycle already, and it's happening extraordinarily rapidly.

The chilling news is not good for Leftist Global Warming fans, who are stocking up on shorts, T-shirts, and sandals for their planned summer street and capital protests. In fact the news is very very serious for everyone:

"The sun is entering one of the deepest Solar minimum of the Space Age," wrote Dr Tony Phillips on 27 Sep 2018.

Sunspots have been absent for most of 2018 and Earth's upper atmosphere is responding, says Phillips, editor of the popular technical site: spaceweather.com.

Data from NASA's TIMED satellite show that the thermosphere (the uppermost layer of air around our planet) **is cooling and shrinking**, literally decreasing the radius of the atmosphere. This is very scary. Because remember that atmosphere, and particularly the greenhouse gasses in it, keep us from freezing.

It's as if you are wearing a nice winter parka on a really cold day, and someone comes and rips away the outer layer.

To help track the latest frightening developments, **Martin Mlynczak** of NASA's Langley Research Center and his colleagues recently introduced the **"Thermosphere Climate Index."**

The Thermosphere Climate Index (TCI) tells how much heat nitric oxide (NO) molecules are dumping into space.

During Solar maximum, when the sun is in a hot cycle, the TCI is high. This means that it will be hot on earth for a long period.

During a Solar minimum the TCI is low. This means it will be cold on earth for a long period.

The following words are far more scary than all the junk on Global Warming that has every been written by the fake scientists. These words of truth are not pleasant news about the present The Thermosphere Climate Index:

"Right now, it is very low indeed ... 10 times smaller than we see during more active phases of the solar cycle," says Mlynczak

Leading NASA and other Scientists Expect Record Cold starting in a matter of months. World leading solar scientists state that a full fledged ice age will hit sometime around 2035. And we will read about that later in this book.

A new ice age! Yikes!

"If current trends continue, it could soon set a Space Age record for cold," said Mlynczak late in 2018.

"We're not there quite yet, but it could happen in a matter of months."

So my dear reader, these very recent developments, plus the clear lowering of the sun's energy which has been noticed since 2009, all shine the light on an truly inconvenient truth.

We are going to become very very cold. And for a very long time.

There have been many ice ages before. During the period of woolly mammoths, those funny looking elephants, there was an ice age.

In all of human history, which is very brief, about 12,000 years, we have only had two ice ages I think. One when humankind was propagating and beating out the competition, and the other, called the "Little Ice Age" just a few hundred years ago during the Middle Ages. From about 1300 to 1800.

This last one wasn't a major ice age, it was a Small Ice Age. But it was cold, and we will take a close look at it in this book to see what it was like.

But before we get to the recent little ice age, let's take a look at the true ice age when the most of the earth was covered by thick sheets of ice. Very very thick sheets of ice, and also mountains made of ice.

WHAT A MAJOR ICE AGE IS LIKE

If you would like to impress someone before we all freeze into little ice cubes, you can use the term "Pleistocene Epoch" when explaining ice ages to them.

That is the time period that began about 2.6 million years ago and lasted until about 11,700 years ago. That was the most recent Big Ice Age. Mountains of ice, and huge sheets of ice, called glaciers and mega-ice sheets, covered most of the earth.

Where I was growing up in Northern Illinois, the area was countryside and it was filled with little valleys called "ravines" (pronounced rah-veens).

They were filled with vegetation and were like little jungle valleys totally different than the flat surrounding countryside.

Ravines were actually dug out as the huge mountains of ice, glaciers, moved forward as things got colder, and then melted to sort of move backwards as things warmed up. So the author of this book, had a growing-up experience indirectly with the last ice age. I used to play a lot in the ravines as a kid.

81

The last glacial period, where mountains and sheets of thick ice were almost everywhere, was between 110,000 and 11,000 years ago. Human beings were moving around during that 12,000 years ago period as things were warming up a bit.

These sheets of ice weren't your ice skating or hockey rink type. Or even the present antarctic type.

They were two miles thick! About ten million square miles in North America and Europe were covered by ice.

The North American Continent was covered in a thick ice sheet on the upper northern half, and Canada was completely covered by ice.

And that is why, dear readers, these periods of time are called "ice ages". There is ice all over the place.

NOW ABOUT THE LITTLE ICE AGE, JUST A FEW HUNDRED YEARS AGO

FROM 1300 TO 1800

The NASA Earth Observatory has pinpointed three recent particularly cold intervals during the time of the Little Ice Age:

One beginning about 1650, another about 1770, and the last ending in 1850, all separated by intervals of slight warming. These occurred in different places on earth and did not cover the entire land area.

The Little Ice Age brought colder winters to parts of Europe and North America.

Farms and villages in the Swiss Alps were destroyed by encroaching glaciers during the mid-1600's.

Canals and rivers in England and Holland, which normally never freeze over, were frequently frozen deeply enough to support ice skating and winter festivals. Those days of people who knew how to enjoy themselves are long gone. The Democrat political correctness has ruined everything.

But winter festivals aside, he winter of 1794–1795 was particularly harsh. And there are many terrible results from such cold weather.

For example, the cold weather, and increased periods in which snow covered the ground, caused famines. In many countries 10% of the population died of starvation due to the impact of the small ice age on farms.

This Small Ice Age was a time of famines, hypothermia, bread riots, and the rise of despotic leaders brutalizing an increasingly hopeless and depressed peasantry. Not so different from the tyrannical Democrat Party of today. But it was much colder then to add to the misery of tyranny.

In the late 1600s during the previous Little Ice Age, agriculture had dropped off dramatically: "Alpine (Swiss) villagers lived on bread made from ground nutshells mixed with barley and oat flour" in order to survive.

One of the key historians writing about this Small Ice Age, Wolfgang Behringer, discovered intensive witch hunting episodes in Europe to agricultural failures.

Witch hunts much like the Mueller investigation became commonplace. Those witch hunts were mobs targeting innocent people. Again similar to the Democrat witch hunts of today.

Then as now there were no witches, but a frightened and desperate population lashed out at anything.

We can expect the same type of awful things to happen at the coming Small Ice Age which is at our doorstep. Famine, collapse of society, violence...

In fact, it will be worse now. Because the rural farming society of the period 1300 – 1800 was better equipped to handle such a disaster. Today everything is mechanized, everything runs on electrical and fossil fuel power which will disappear during an ice age.

The Global Warming maniacs are going to get their wish, but not as they had hoped. The coming Ice Age, depending on its severity, could destroy the ability to secure and use fossil fuels. Alternative energy systems such as wind and solar won't work either in an ice age. That is covered later in this book.

In even a Small Ice Age, almost everything we depend on, collapses from the cold. Elevators in buildings, for example won't work.

Either because they are frozen or there is no electricity.

No water pumped to homes and apartment either.

Witch hunts then happen as the society collapses. It doesn't take much imagination to see that Al Gore, the Science Guy, and the Democrat Party generally, will not want to be found by the mobs at that time.

Now among the most unpleasant news about what awaits us in the soon-coming new little ice age, or even major ice age, is that societies deteriorate as the result of the catastrophes caused by the cold.

These catastrophes are going to include in America: lack of food. Growing seasons are shorter, and it's going to be too cold to grow many crops.

Not only the food raised from the ground (heavily reduced crop growth), but milk, livestock and so on collapsed in the previous little ice age. And will happen in the coming one, just around the corner.

In addition there was lack of clean water.

Disease was rampant.

Hypothermia deaths from exposure to the cold.

Mas unemployment and so on.

In the previous Little Ice Age which ravaged North America and Europe just a few hundred years ago, disease and unemployment fed off each other.

And the societies degenerated into violent attacks against scapegoated groups which had nothing to do with the climate.

"Witches" were the number one target. Prior to the little ice ages, not much attention was paid to witchcraft. But as societies deteriorated, elderly women, widows, or any woman were an easy target.

Generally speaking, violence escalated in society, against women, against anyone. Sexual crimes were rampant, ranging from rape to bestiality.

Theft was rampant. Murder was rampant.

In short, the name "Little Ice Age", sounds like a charming little change of climate. But it was anything but that.

And today we can expect the same things to happen. Probably a lot worse.

One of the dangers of the Global Warming & Climate Change hysteria, is that it is taking massive resources to solve a problem that doesn't exist.

The Global Warming / Climate Change From Human Activity fanatics are pushing governments, peoples, in the wrong direction. So we can not properly prepare for the coming ice age.

The responsibility for the coming disaster is on them. They didn't make the coming frozen weather, but they are keeping us from preparing for the coming disaster.

The Global Warming / Climate Change Caused By Human Activity mob is yelling about a hot future with floods along all coastlines.

Instead we are heading over the cliff into an Ice Age.

CHAPTER 14

BUT WHAT ABOUT THE LATEST SCARY UN OR GOVERNMENT PAPER ON GLOBAL WARMING...

AS USUAL, BIG, REALLY BIG MATHEMATICAL AND CONCLUSION MISTAKES IN IT...

I WONDER WHY?

The United Nations is always cranking out scary papers on Global Warming. Since the UN can't do anything right, they have to try to find something, anything to look relevant and good. So they picked fake Global Warming religion. It's called "virtue signaling". It's doing something to make them look good at something, although they're not.

When the UN is not supporting terrorist states, such as Iran, or human rights violators such as Turkey, Sudan and so on, they busy themselves by supporting Global Warming. What could go wrong with that?

In the late 1980s the U.N. was already claiming the world had only ten years to solve global warming problems or face the consequences of disaster.

For example, the Left wing newspaper, the San Jose Mercury News, reported in June 1989, that "senior environmental official at the United Nations, Noel Brown, says entire nations could be wiped off the face of the earth by rising sea levels if global warming is not reversed by the year 2000."

That prediction obviously did not come true 19 years ago. Except for Noel Brown, who reportedly in trying to avoid rising sea levels moved to the Sahara Desert in Africa. And died of thirst.

That's just one of the hundreds of deadlines of disaster that Global Warming will supposedly cause according to the UN. They are all proven wrong. All fakes.

And sadly today the U.N. is still sounding the same false alarm today to a newer and younger group of Global Warming religionists.

Young people don't know about all the fake and scary predictions in the past that never happened because these mistakes are kept hidden.

Others simply forget all the prophesies of doom because there are so many of them, that never happen.

Al Gore has been saying the world will end from Global Warming every year since 1990, and it never does. OK, now this book is telling readers on the Left about the failed hysterical predictions.

But let's get up to date and let's look at some of the recent Global Warming crazy reports.

Let's suppose you are a tourist, visiting New York City. And you buy a guide book about the city, so you can learn about it, and find your way around.

So far so good.

But when you open the book it tells you some very strange and very incorrect things. Such as that the population of New York City is 850 people. Yikes.

The real number is something like eight million five-hundred thousand people. A mathematical error just like in Global Warming always.

So your guide book says:
850 people

But the reality is:
8,500,000 people

You're upset and you call the publisher. And the publisher tells you, don't be upset. It's because we made a mathematical mistake. The person working on the book for us didn't know how to do math. So don't worry about a thing.

But you're still upset of course. The silly book "Get to Know New York Without Taking the Time to Discover the Truth" cost $39.95!

Anyway, you want to meet some friends who live in New York, so you open your new guide book and go to the map section of the subways.

Specifically, you want to go to Grand Central Station to meet your friends, which is in the center of the city.....

But by following the model of the subways in this book, you end up in lovely Lebbus Woods, a massive

87

slum area,eighteen miles away instead.

You miss your appointment and your friends give up on you.

So you call the publisher of the book, "Get to Know New York Without Taking the Time to Discover the Truth", and tell them in a loud voice what happened.

So the publisher, which is actually our model for the UN, tells us...."Oh don't worry about a thing! It's just that the computer model we used to show the subway system doesn't work. It shows everything backwards!"

So you say, "Can I have my money back?"

"No way! We didn't do anything wrong!", and they hang up.

Well guess what? The newest of the UN papers on Global Warming is just like that book you bought.

Only it didn't cost $39.95. It cost $39.950,000 (thirty-nine million dollars). And no refunds available either !

Basically you can assume that these UN reports are bunk. So far, every single one of them is. And the climate models on which they are based, have bene proven to be falsely predictive of the future. And completely wrong about the present.

In an open letter to an earlier UN report, one of America's leading and most famous meteorologists, James Coleman, wrote this in an open letter to the UN Climate Change committee authors:

"The ocean is not rising significantly.

"The polar ice is increasing, not melting away. Polar Bears are increasing in number.

"Heat waves have actually diminished, not increased. There is not an uptick in the number or strength of storms (in fact storms are diminishing).

"I have studied this topic seriously for years. It has become a political and environment agenda item, but the science is not valid."

"There is no significant man-made global warming at this time, there has been none in the past and there is no reason to fear any in the future.

"Efforts to prove the theory that carbon dioxide is a significant greenhouse gas and pollutant causing significant warming or weather effects have failed.

"There has been no warming over 18 years."

In other words, Mr. Coleman correctly told the UN global warming group based on data, to shove it.

Pardon the harsh words, but that's what real scientists are telling the UN. Even many top scientists who are members of their own climate committee have abandoned ship and are saying these things.

But because funds for doing "new research" if you lie, are flowing like a river, every now and then there come additional "prestigious" studies spreading fear and terror about the non-existing Global Warming menace. Hysteria is an objective of these new climate reports.

There is no global warming, and no climate change caused by human activity. So the human-cased global warming and climate change they report, exist only in the minds and pocketbooks (they get paid a whole lot for writing their garbage) of the authors.

The most recent of these studies was to be officially published, quite rightly on the date of October 31, 2018.

Making it a Halloween study. A time for imaginary monsters. This study fit the holiday of its birth perfectly.

And perfect fit for the Global Warming / Climate change fraud. Because what else is the fraud except something to scare you with fake costumes.

Only instead of witches and goblins with real people inside, we have scientists who are apparently, and with all due respect, actually witches and goblins.

We have already discussed this study... the Global Warming / Climate change Halloween study study was published in NATURE journal, which used to be a well respected... before NATURE got in to Global Warming publishing...

Quantification of ocean heat uptake from changes in atmospheric O2 and CO2 composition

And yes, We haven't mentioned previously, that this study was published on Halloween.

It appears that NATURE was in such a big hurry to publish this Halloween science article, they didn't review the study carefully before turning on the printing machine. Oh well, anything to move Global Warming narrative forward, true or not.

We talked about this study before in this book, but let's take a closer look at it. It's the poster boy for Global Warming fraud errors.

Now "Quantification" in the study, means in the distinguished authors' impressive academic language, to count something.

For example, when we ordinary people (not the 10 great scientists who wrote this) go to the grocery store we buy 3 gallons of milk (if we like milk), 1 loaf of bread, 5 avocados, 6 candy bars, 1 box of dogie treats for our dog, and so on. Counting.

So for this study, the ten prestigious university researchers counted how many degrees of heat the oceans took into the water from increases of Oxygen and Carbon Dioxide in the air around us. Very exciting stuff.

And the results were really scary. As would be suitable for its publication date on Halloween... It seemed as if Global Warming was much worse than we thought! And that the world was certainly heading for disaster from Global Warming.

We didn't see the heat, claimed the study, because the heat caused by Global Warming existed, but it was hiding in the oceans!

The Global Warming movement licked its chops or possibly clicked their heels if they felt athletic.

THE DEMOCRAT MEDIA COMPLEX REACTS

As usual, when this report was published, Leftist political organizations and media expressed happy hysteria, that their hopes had been fulfilled. Finally they had absolute proof that Global Warming was real and very very dangerous.

They took off in glorious flight of victory. Global Warming had been proven at last:

But the victory of this article for the Global Warming mob was in one sense bittersweet. The Leftist political organizations and media were a tiny bit sad and gloriously happy at the same time.

They were a tiny bit sad that the world was being destroyed by Global Warming, but they were even more happy that they had been right! That's all that really matters to them.

Money and power is the game.

And that now they could get more power and money than ever to control the economy and the lives of ordinary people. You know, control the deplorables.

As the result of the Halloween paper, victory of the Democrat Party, Justice Socialist party was in sight.

After the paper was published in early November, 2018, the Washington Post, the New York times, had **orgasms of joy**, as they read in the report that Global Warming is even worse than they thought.

And that the world is now near a point of no return in which the planet would be destroyed by heat from CO2 which we caused ourselves! Yikes!

"See! We told you so!" parroted the Democrat Media Complex, and the pseudo climate experts. "Global Warming is real and destroying the planet!"

The Washington Post called the climate study **"startling"**!

And their fellow travelers at the New York Times said with glee, that the study suggested global warming "has been more closely in line with scientists' **"worst-case scenarios."**

Oh my!

The European version of the American Global Warming media hysteria, the BBC, based on this study, proclaimed to its viewers somberly

"This could make it much more difficult to keep global warming within safe levels this century."

Oh no!

YOU CAN GET YOUR VERY OWN COPY OF THIS STUDY !

And now ladies and gentlemen, and children of all ages, all the Global Warming fans who would like their very own copy of this historic scientific article can buy it for about $15 or $20 from the journal website, nature.com.

It's really a fun article to have, and you can put it in a picture frame on your wall. Or even paper walls of a small room with it.

The Gang of 10 Global Warming warriors who wrote the paper, estimated ocean heat by measuring (as we mentioned previously) the volume of carbon dioxide and oxygen in the atmosphere.

The shocking, horrifying, incredible results: the oceans took up 60 percent more heat than previously thought!

The study sent alarm bells ringing, especially coming right after the latest alarmist UN Panel Climate Change panel study which itself proclaimed doom and gloom.

"The planet warmed more than we thought. It was hidden from us just because we didn't sample it right. But it was there. It was in the ocean already," Resplandy told The Washington Post at the end of October.

The political Left was foaming at the mouth with excitement and joy. Most of them bought a copy of the article, even though they couldn't understand it. They put it perhaps in their pillows. Or made underwear out of it, to keep it close to their hearts.

But hold on to your $15. Perhaps there's no reason for anyone spend $15 or $20 or even 15 or 20 cents, on buying a copy of the study, for the following reason.

As we mentioned, there was a little teeny weeny tiny problem with the study.

And that little teeny weeny tiny problem was with the "Quantification", the math.

Ho hum... Global Warming / Climate Change science as usual...

Chinese scientific commentators said it was like the Gang of Ten global warming scientists went to the store for six candy bars but couldn't count them right and came back with 60 of them. The results of their study was like that.

Happily, for all of us who love truth, and dear reader, and by now every one of us is part of that group....

a mathematician, and one of America's leading climate scientists, having many brilliant published papers, Nic Lewis found the new paper's gloom and doom findings...

all stemmed from a math error.

Lewis said "a quick review of the first page of the paper was sufficient to raise doubts as to the accuracy of its results."

After correcting for the math error, Lewis found the paper's ocean warming rate "is about average compared with the other estimates they showed, and below the average for 1993–2016."

Yikes, Laure and her nine little friends bought 60 candy bars instead of the intended 6!

Lewis also criticized the climate model predictions, which were part of the study, which generally over-predict warming.

Lewis' critical results of the "study" were replicated quickly by University of Colorado professor Roger Pielke, Jr., who shared his analysis using Twitter a few days later.

Professor Pielke confirmed the linear trends reported by Resplandy and her colleagues stemmed from "a big error at the core of paper's findings."

The Global Warming / Climate Change so-called "scientists" cause far more damage with their lies than loose wheels or a poorly constructed roof.

The work of the Global Warming / Climate Change movement are leading the country to disaster on a highway of lies to national suicide.

I wonder where this math illness comes from? Did ten university professors and researchers suddenly forget their high school math? Was it something they all ate?

Although their so-called break-through research was fake, it fooled the New York Times, Washington Post and the whole Global Warming movement.

Because there is a Left wing political agenda. A political agenda of big government run by a small self-proclaimed "elite" of Democrat media slime which lords over the people.

Big government, heavy taxes, societal collapse with open borders, it's a dream the Left can accomplish by Global Warming lies. The lies are intended to bring hysteria in the population, that something be done at any cost.

Done at any financial cost, and done at the cost of freedom.

Global Warming / Climate Change is a political movement, not a science. It's also a religion not a science.

But it's not impossible to escape it. It's not impossible to give it up. It's not smoking or drugs. It's a delusion.

IT'S POSSIBLE TO BREAK FREE OF THE BRAINWASHING AND DELUSION WHICH IS GLOBAL WARMING...

As an example that anyone can get free of the slavery of the Global Warming hoax, we have in fact the father of Global Warming. He is one of America's most well-known scientists. James Lovelock.

In 2006 he wrote the classic book on Global Warming. Warning us about burning doom in the near future.

But just like all the Global Warming literature, it was quite wrong. And 20 years later he admitted he was totally wrong. Here is what he said to the BBC recently.

He slammed global warming claims including those of the United Nations Climate Panel:

'They just guess. And a whole group of them meet together and encourage each others' guesses.'

And Dr. Lovelock, the scientific father of the viper called Global Warming, now calls global warming "a religion, not science":

"IPCC (the UN Climate Panel) is too politicized & too internalized' -- 'I don't think people have noticed that, but it's got all the sort of terms that religions use.

It's becoming a religion, and religions don't worry too much about facts."

When asked recently how it is possible he was all in for Global Warming but now is totally against it as a false idea and a new religion, he said:

"Well, that's my privilege. You see, I'm an independent scientist. I'm not funded by some government department or commercial body or anything like that. If I make a mistake, then I can go public with it. And you have to, because it is only by making mistakes that you can move ahead."

Lovelock dismissed the entire basis for global warming concerns in his BBC television interview.

"Take this climate matter everybody is thinking about. They all talk, they pass laws, they do things, as if they knew what was happening.

I don't think anybody really knows what's happening. They just guess. And a whole group of them meet together and encourage each other's guesses," Lovelock explained.

A UN Lead UN IPCC Author, Dr. Richard Tol, joined Lovelock very recently in slamming the UN and had his name removed from the (United Nations) International Panel on Climate Change report.

And he is only one of many who have abandoned the sinking UN Global Warming / Climate Change ship.

So dear Republicans, you can let your politically Left friends and family know, in a loving way, that anything they read in support of Global Warming is a lie.

But there is an escape hatch. They just have to say.. "I am free."

Regarding the magical and mystical Halloween computational study we have discussed... basically all the Global Warming / Climate Change research is like that. Either by intent or by error, it's wrong. And unfortunately it's leading many on the pathway to national suicide. An Ice Age is at the doorstep. Scheduled for arrival around 2025 or earlier.

No one wants to base their beliefs and actions on a lie. But everyone who supports the Climate Change / Global Warming movement is doing it.

Well, I shouldn't say "no one" wants to base their beliefs and actions on a lie. It seems to be an illness among all Democrat Party, Justice Demcrats, and Democratic Socialists.

In November of 2018, after AOC won her election race in completely Democrat New York City district 15... and after showing up in Washington DC to start doing strange things, she managed to stop into a protest which was taking place in the office of none other than Nancy Pelosi.

The 40 or so Global Warming supporters who were there, were protesting on a workday. So we can assume they were unemployed or maybe they are employed by Soros or Steyer, or Bezos, or Bloomberg "environmental" organizations.

Anyway, AOC, the inspiring leader that she is, told the protestors sitting on the cold floor of Nancy Pelosi's office them that she was proud of them.

Paraphrase: "I'm proud of you. You are acting to save the planet!"

Sorry Alexandra, but you got it wrong. Your little friends are doing things to destroy the planet, not save it.

We are heading into frozen hell, and you and your protest buddies are doing everything possible to keep us from getting ready for it.

Dear reader, Vote Republican. As if you life depended on it. Because it does. And no votes for RINOs, just for solid constitutionalists conservatives... no more Mitt Romneys!

Do everything you can that is within the law, to fight against the coming Democrat Party efforts in the House of Representatives, and their mind-numbed robotic protesters who believe in global warming.

The Democrat Party is delusional, power-hungry and they are driving the country into national suicide.

In addition to voting, we need to do everything possible to stop Democrat vote fraud.

Democrat vote fraud is massive in all US elections. University studies estimate around 2.8 million illegal alien votes are cast nation-wide in a presidential election.

So we need to stop that now. We need to use their mantra, "Count every vote." because for us it means that every false vote, every illegal alien vote, negates the vote of a registered American citizen.

Our legal votes are crossed out by their illegal ones. So we have to stop Democrat election fraud, as well as to be sure that we ourselves vote.

The future of the planet and our families depends on it.

CHAPTER 15

THE NEW ICE AGE IS COMING SOON

DOES THE AUTHOR OF THIS BOOK ACTUALLY KNOW SOMETHING ABOUT COLD WEATHER?
WELL..... YES

As the author of this book, it has been my pleasure to be making this journey with you through Climate Change, the Halloween of science.

I thought it might be nice if I told you a little about my work experiences with cold weather. Very cold weather.

First, I grew up in a semi-rural area of Northern Illinois. The place where as is mentioned previously in our book, was filled with small valleys called "ravines".

The were a completely different place than the rest of the countryside. They all had little streams running through them and were sort of northern hemisphere tiny jungles. Anyway, as a child they seemed like jungles to me and I loved playing in them without end.

It's good my parents didn't know much about it, because those paces were probably dangerous. Easy to get lost for small children. And if something happened, it would be very difficult for anyone to find you.

In fact I had a little friend, I think we were 10 years old. And one day he went out from home and apparently just disappeared.

Thanks to his dog which was a red haired Irish Setter, that was with him and kept barking, he was found in one of the ravines. It was even in the local newspaper with a picture of his dog by his side after returning home. But that experience of a friend didn't stop me from playing in the ravines.

What I didn't know until I got a bit older, and why I mention ravines in this book, is that the ravines were caused by glaciers, by sheets of ice. As the ice sheet spread slowly from North to South, it dug these little ravine valleys with its huge weight.

And in the warming period after the Ice Age of that region, the ice sheets and the glaciers (little mountains of ice) receded.

So I grew up in a former glacier and ice sheet-filled area.

Now the ice sheets of the massive ice age 12,000 years ago, were not like that of a hockey rink or ice skating rink. They were up to one or two miles thick! It was a tough environment during major ice ages.

Winters where I lived as a child were cold and snowy. And we wore ear-muffs, heavy mittens and lots of clothes.

But that's not the extent of my cold weather experience and studies.

After I later became an engineer, I got a very good job in Scandinavia and so I went there. Actually I went there to give a summer seminar at a university, and liked the place so much that I stayed on in good job doing alternative energy consulting.

I am a believer in renewable energy and work in the solar thermal field. I also understand that alternative energy (meaning the wind and solar the Left loves so much) can provide perhaps a maximum of 12% of our energy needs. And that's under optimal conditions. Very hot, very cold means it's not going to be optimal.

Without fossil fuels and the new excellent low pollution coal technologies, we as a society and country are finished. Back to the stone age.

And that's particularly true, because a major source of US electrical power comes from nuclear plants. 20%.

But the US mines only 2% of the uranium needed to power the plants. So we are in real trouble regarding nuclear power, and our only hope to make it through the coming colder periods is fossil fuels.

Anyway, I thought my consulting job in the frozen north would be for a year or so but it turned out much longer.

And the longer I stayed, the more I started to like to do research in the very cold climate.

The weather in northern Scandinavia is chilly to say the least. Many winters where I lived, it reached -30 degrees C often during the winter. In our US Fahrenheit system that's a cold -22 degrees.

I used to like it, because I had the right clothes. Nice warm boots, special ski pants, big fat Scandinavian gloves that looked like white boxing gloves, and so on.

I could march around a few hours in the -22 F. But that was enough for me at one time. It stopped being fun after two hours.

Scandinavian people could stand it more than I could. By-the-way they are very nice people as well as very tough when it comes to cold weather.

Still I went up north sometimes to do cold weather research, and it got down to -45 degrees C (which is in our Fahrenheit system is -49 degrees). So I've been working outside often in really cold weather.

I remember when my older brother was in the army, and he was in Alaska. I think he said that it got down to -60. I don't know if that was true or not, but it interested me.

Evidently my experience of -49 degrees Fahrenheit wasn't cold enough for me. So when I got the chance to do a six month renewable energy engineering project in Havöysund, Norway, I jumped at the chance.

Havöysund is not the coldest place I've done research, in fact it's not even close, but it is so far north that it's like stepping into another world.

The people in the village were wonderful, and the village itself was very nice. But step outside the village and you get an idea of what the Ice Age World will be like.

It's not just the snow and ice, its a very very spooky feeling.

Here it is in the in the most lovely part of the summer:

It's easy to get a sense there in the northern most village in the world, how big the planet is and how tiny we are as humans are. And we are tiny even as countries, in addition to tiny as individuals.

When winter comes in Havöysund, it's something like an ice age. Outside the beautiful little village, the place in the wilderness is absolutely scary if you're not used to it.

Even my many years in northern Scandinavia at -49 degrees did not prepare me for what ice ages are like.

Here is a place on the general area outside the village in the springtime. It's a vast emptiness.

As long as I was in the village, everything was OK. Even when I would get up in the morning and find that the snow had completely covered my front door with two yards of new snow and I had to dig myself out. That was OK. I was used to lots of snow from previous work.

But when I walked out of the village things could get a little uncomfortable. From that experience I got a feeling of what an Ice Age might be like for us.... until we all freeze that is if it's a major ice age and not just a Little Ice Age.

As I mentioned, the village was very nice and the people were also. Amazing fishing with giant fish for tourists and the fishing industry in the summer.

But getting a bit outside the village and the size and depth of the cold nothingness hits you.

It becomes very clear that our planet is very big, massive and powerful. And that we are something like ants. It's a feeling that there is no control by us possible. That natural forces which control climate (meaning the sun) are much more powerful than humans and are going to do what-so-ever they want.

I occasionally walked up to the series of wind turbines on the little mountains around Havöysund. On these high hills surrounding the village there is a new high technology alternative energy system of the type which the crazy Green New Deal proposes as a way to power the country.

There is a good lesson for the Green New Deal delusional Democrats, Justice Democrats / Democratic Socialists from Havöysund. The reason you can walk right up to these huge wind turbines in the Autumn and Winter is because the huge wind turbine blades are not turning.

Some investors from farther south, I think Denmark, believed Havöysund would be a great place to generate electricity and then send it by wires down south a bit. They were the first Green New Deal.

But the spooky sea spray in the air settled on the turbines and froze. And in any case the turbines themselves froze. So in the very long winter, the whole system is just a bunch of huge white statues against a gray sky. They don't move. They don't produce electricity.

The wind turbines are frozen shut.

In the photo here, the blades are still not turning, and summer is just around the corner.

The snow there looks gray also because there isn't much sun and the sky is gray. Maybe the grayness also stops the blades from moving. But they do not turn.

To be fair, I have to say that the wind turbine blades manage to turn a few times during the summer. But they have been essentially ruined as a reasonable power producer by the frozen environment.

AS WE MENTIONED PREVIOUSLY, GREEN ENERGY DOES NOT WORK IN A CLIMATE CRISIS

WHEN IT GETS VERY COLD, OR VERY HOT, WE CAN KISS ALTERNATIVE ENERGY EQUIPMENT GOODBYE

And here in Havöysund we can see how the cold weather has crippled the wind turbines.

I was in Havösund to do solar thermal R&D. When they found out I was there, they asked me if I had any technologies to help get their frozen wind turbine moving in the winter. I didn't and nobody else did either.

For those depending on the electricity which isn't coming to Denmark, it's a feeling of helplessness. And that fits that amazing huge environment. Which I believe is similar to that of an Ice Age.

The feeling one gets in the open country outside of the village is this. It's as if a voice is telling you: "The earth is big. You are not." There is or course no voice, no sound at all. And the feeling is: we've
been had by nature.

If AOC goes up there, she will understand that there is more to worry about than her garbage disposal.

Dear reader, this chapter is just a little personal note, to let you know your author has actually been there. And in places for long period of time for research on infrastructure designs for ultra-cold climate, much much colder than Havöysund.

But we can't get ready, because the false religion of Global Warming and Climate Change being caused by human activity, is tying our hands.

Oh if Global Warming were true, how happy we all could be! How happy the world would be.

But it isn't. It's only a fraud and is keeping us from getting ready from what is on the doorstep: a very scary Ice Age.

CHAPTER 16

TOUCHY-FEELY PROOF A NEW ICE AGE IS COMING VERY SOON

One of my friends, who is a political Leftist told me something as I was writing this book. She said, that to demonstrate to people who believe in Global Warming and Climate Change that it's a fraud, the book had to include some strong touchy-feely proof that an ice age is coming.

"Sometimes facts don't matter to us on the Left", she said, We're more into it emotionally."

OK, so this chapter is dedicated to my friend's advice. Truth yet presented in a more personal way. And bringing in a very important and interesting long-term friend of mine who is Native American. And who is an expert on things of nature.

To get expert advice in writing this chapter, I called a very wise friend that I have had for many many years. As mentioned, he is Native American, in touch with the environment and really has a lot of wisdom. From directly with nature, and from study, and massive amounts of reading. He is also a solid Republican.

I now explain how we met, and give some background about my wonderful friend.

I grew up in a semi-rural area of northern Illinois. Every summer, from the time I was six until I was 18, I spent on my uncle's farm.

There I learned how to do many things, including how to make butter, ice cream, shell peas, collect eggs, and once even clean a chicken that had been slaughtered for meat (I only did that once... although I was invited to do it again).

Anyway, in the wooded areas where I lived, I occasionally found arrowheads. I think I was about seven or eight years old the first time I found them.

That got me very interested in Native Americans. I collected many things from areas in which they hunted, mostly arrowheads, but one or two tomahawk stones as well. I read a lot about their way of life. And had a few picture books about their way of life in nature.

So I became very interested to meet Native Americans and to become friends with them.

I told my parents about this, and once on a trip to Florida when I was 8, they made sure we visited a Seminole Native American reservation, through an invitation from the Native Americans.

On that trip I met a young boy my own age. After spent the day playing together, we decided to become pen pals. And have been in contact off-and-on, ever since. I used to have a photo of us together while I was there, but it's long faded.

So in writing this book, I decided to call him and ask him if he thought an ice age was coming. He certainly knew some things I don't know, according to his many skills in nature, and from the vast amount of study and reading he does.

I called him and asked him if it was OK for me to ask him a question about the weather, and he said it was OK.

"Do you thing it's going to get a lot colder? That an ice age might be coming?"

Usually he does not talk a lot. Sometimes I would write him a letter, or more recently e-mails, and tell him how everything was with my family, and ask him how they all were doing. Mostly he would answer briefly, "Praise the Lord."

Anyway, he started talking a lot, after I asked him the question about a possible ice age.

At first he talked about everything but climate.

He told me how nice thing were in south Florida now, in the reservation area, after Donald Trump had become President. He said that, "jobs have popped up all over the place. Young people who were formerly unemployed, and who would get into trouble, now could get a job easily."

"Businesses which were started by our people, and which were doing poorly under Obama, have suddenly become profitable. Less regulations, less taxes. They are growing and hiring more people."

"More young people are heading to college and university also. So things are doing very well here."

"We've had a bunch of big food shops open and restaurants open too. Which brings down our cost of living."

I wasn't sure why he was telling me all this, but then he said:

"Joel, you better get your family down here to Florida soon. Get your behinds down here soon, and I'm not joking. You are going to freeze and die in Chicago."

I reminded him that I moved from Chicago to South Carolina, but he still said, I had to move to Florida, because everything north of Florida was going to get very cold.

He explained:

"You know I have read a lot, and listened to our elders, about the history of my people, written and told by my own people. They said we came to this North America from a very cold place, which on the map, now is written 'Tibet', and which is like an ice kingdom."

"I know that is true because I have seen many photographs of the Tibetan people and they look exactly like us. No differences. And they are tough like us."

"I saw paintings showing Native Americans about 12,000 years ago coming from Tibet to America over a land bridge or ice bridge. And there were huge prehistoric elephants called mammoths in the background."

"Now here is what I think and believe, so I can answer your question. I did not forget what you asked."

"My people, and in fact most Native Americans have suffered a lot from the governments of the US. We used to roam free but now we are on reservations and have been on them for a long time. In the 1800's we were told to trust the 'Great White Father' which meant the President of the United States. But I and my people never found them to be great or a father. Except for Abraham Lincoln and a few more Republicans."

"Lincoln freed the slaves. He was a Republican, and we thought that next the Republicans would free us too. And that we could roam freely again around our lands."

"But the Democrat Party, the same one what was for Slavery and for the KKK, kept us down. The Democrat Party controlled the South where we live."

"Things started to change for the better a few years ago when more and more Republicans came to office in the South and including Florida. Things have been getting better and better for us under Republican rule.

"That is why I always vote for the Republican men and Republican women too. And encourage our younger people to get involved in politics and become candidates. Some have and are in local office. Some are in the state government."

"Anyway, about what those slavery and KKK type people, all the awful things the Democrat Party did to us, a great punishment will come upon their own area. Not on our reservations, but on their own area. A great wind from Tibet, which will freeze them."

"You see Joel, the government put us in the hottest places in the USA. Thinking it was a sort of punishment. They put us in deserts and in dry places. But now it backfires on them."

The backfire started maybe 50 or 60 years ago, when the Agua Caliente Native Americans in southern California, were living in poverty in their hot desert area they are restricted to. It was just dry desert with nothing in it. So hot."

"It was not their native homeland, but a desert punishment. However, guess what? Uranium was found on their land, and they all became millionaires. It shows you that the Lord is helping Native Americans."

"Many casinos, as well as hotels and other native American-owned businesses also have brought wealth to reservations."

"And soon will come the cold wind from Tibet, our former frozen home, which will turn the entire US, except for the hottest places, which is where we live, into ice."

"Therefore my old friend, my blood brother, I urge you to being your family here. I will arrange with the elders for you to be able to rent a nice place for your family. With all the jobs and economic activities and new shops under President Trump, it's a great place to live."

That was more talking that he usually does. It was important. And I was glad to hear he had been kind enough to arrange a way out for my family.

I was quite happy to get the invitation, because I've been wondering where the best place to live might be after the cold of the Ice Age starts.

So for our readers on the political Left, maybe you can't accept the information, the facts provided in the previous chapters... but many of you can find some connection with what I just said.

And or course, the Republicans, which are the true party of freedom. The party of the future. The real party of compassion and caring, can too get ready.

Unless the water, transportation, and food infrastructure in the US is upgraded, and fast, for ultra-cold weather, we are in for a very tough and unpleasant time.

CHAPTER 17

THE DEMOCRAT MEDIA COMPLEX:
BRAINWASHING THE COUNTRY ABOUT GLOBAL WARMING

KEEPING US FROM GETTING PREPARED FOR THE REAL DISASTER

The climate war on America, the Global Warming fraud, is coming from several sources.

One is of course the political base. The Democrat Party. And the dark totalitarian grab for absolute power.

Another is the financial war, which will later in this book be covered in detail. In which billionaires, lawyers, alternative energy companies, and multi-millionaires become more wealthy through the Global Warming scam.

And a third is the Democrat Media Complex.

This includes but is not limited to the Global Warming activists who are totally ignorant about climate and solar effect on the planet's surface temperature, but think of themselves as experts anyway:

- **The Washington Post and the Holocaust-denying New York Times**
- **MSNBC, CNN**
- **ABC, CBS, NBC**
- **The White House Press Corps**
- **National Public Radio, National Public Broadcasting, Bloomberg Media**
- **Internet garbage sites such as Buzzfeed, Politico, Huff Post, Mother Jones and many others**

They keep a steady drumbeat of lies, and attack anyone who dares to contradict the Global Warming / Climate Change fraud.

n April 29, 2019, an excellent article published here defined three groups of scientists working in the global warming field. The Three Sides Of Climate Science, by Hans Schreuder, defined three groups of climate scientists.

• Those who are all in, and hold the theory that CO2 is destroying the planet.

• A second group called the "lukewarm" group holds the theory that CO2 is warming the planet, but not very much.

• The third group, rational, sane, independent scientists who hold the accurate position that CO2 has no effect on global warming.

In this present article, we spotlight the fourth group of self-proclaimed scientists: the Democrat Media Complex.

Membership of this elite group of new climate fraud scientists includes but is not limited to the above mentioned group of media companies.

AND IN THIS DEMOCRAT MEDIA COMPLEX BROUP, IS ANOTHER GROUP OF SCIENTISTS, A STRANGE NEW ASSORTMENT...

MEDIA PEOPLE WHO HAVE BECOME CLIMATE ACTIVISTS WHO SEEM TO KNOW EVERYTHING...

Chief scientist of this new crop is apparently Chuck Todd of NBC, MSNBC.

His name used to be "F. Chuck Todd", but somehow, he managed to summon the mental resources to make it to the courthouse to have his name changed. He apparently didn't realize that the "F" was the reason people trusted him. And now it's gone.

Anyway, Chief Climate Scientist Chuck Todd issued a science proclamation that aired his television program Meet the Press December 30, 2018.

Chuck informed the nation, and in fact the world:

> We're going to take an in-depth look ... at a literally Earth-changing subject that doesn't get talked about this thoroughly — on television news, at least — climate change."
>
> But just as important as what we are going to do this hour is what we're not going to do. We're not going to debate climate change, the existence of it. The Earth is getting hotter. And human activity is a major cause, period.

> We're not going to give time to climate deniers. The science is settled, even if political opinion is not.

Well then, we can all relax. It's settled.

But perhaps Chief Climate Scientist Chuck made a tiny error in his introduction. When he stated that "climate change" is a subject little discussed on the media.

In fact, the Democrat Media Complex is obsessed with promoting the human-caused global warming scheme. They live and breathe it.

They not only flood their media with people from Science Group One (those who are all in that CO2 is destroying the planet and sending us into a man-made oven) ... but their own "reporters" have morphed into scientists.

As for the printed word, the anti-Semitic New York Times and the truth-crushing Amazon/Washington Post are particularly active in headlining every study backed by Michael Bloomberg, Tom Steyer, The Sierra Club, George Soros, the UN IPCC, and all the other masters of deceit.

These print media outlets comment on studies as if the reporters actually knew how to read them.

THE POWER OF THE PRESS

The Democrat Media Complex is the focus of this article because the so-called "press" is a major power behind the global warming fraud movement.

Without the Democrat mouthpieces and hacks in the media, the BIG LIE of "global warming caused by human behavior" would not have grown as it has.

The Democrat Media Complex is evidently hell-bent on fulfilling its mission as Enemy of the People.

But what does it really mean "Enemy of the People"?

The term does not originate from Stalin, as the Fourth Group of Climate Scientists, the Fake News Media, would have us believe.

It was coined in a Norwegian play written by Henrik Ibsen in 1882. Called of course, "An Enemy Of the People".

In this play, a Norwegian town of the 1880s, which is dependent on its spa waters health center for economic life, discovers through its city doctor and government testing that the water is heavily contaminated and it poses a serious health hazard.

This disastrous news would certainly mean the end of the city. And the end of its upper economic and social class that is dependent on the spa for its economic and political power.

Now here's the rub: The town newspaper has the information that the spa water is toxic, but prints articles that the waters have been given a clean bill of health and the spa waters are safe. In other words, the newspaper prints the BIG LIE.

And doesn't that term "Enemy of the People" fit the Fourth Group of Scientists, The Democrat Media Complex, perfectly?

To that end, here is a message which is directed at the Fourth Group of Climate Scientists, led by Chuck Todd. It's a quotation by UN IPCC's Japanese Scientist Dr. Kiminori Itoh, an award-winning Ph.D. and environmental physical chemist:

Warming fears are the "worst scientific scandal in the history...When people come to know what the truth is, they will feel deceived by science and scientists."

Yet, this and other statements by real scientists will fail to sway the Democrat's media machine.

Like ancient anti-science figures, the Democrat Media "reporters" are blind. And leading their culture and society into a downward spiral of intellectual disaster... dragging the country into fire, death, and destruction.

CHAPTER 18

OUR PUSHBACK AGAINST THE DEMOCRAT MEDIA COMPLEX IN THE BATTLE FOR TRUTH

FREE FOR THE READERS OF THIS BOOK, OUR VERY OWN NEW AND AMAZING VIDEO

JUST GO TO THE YOUTUBE LINK, AND SEE THIS AMAZING VIDEO ABOUT GLOBAL WARMING FRAUD, UNLIKE ANY OTHER YOU HAVE SEEN BEFORE

We have just discussed the Democrat Mainstream Media. Working as a powerful force to brainwash the American people.

Not to be left behind, the face of the Green New Deal, AOC, has her very own video.

Global Warming / Climate Change Caused by Human Activity are in the news often.

But the Ice Age is not in the news, it's not in the Democrat Media Complex at all.

The Ice Age is not there because the Democrat Media Complex is not about news, it's a propaganda mouthpiece for the Democrat Party.

Working to assist them to seize power. And to do this through lies and through censorship of the truth.

Right now the breaking news, or at least a hot topic, is the so-called **Green New Deal**.

A better name would be Green New Steal.

Because it's all about stealing money and stealing power. It has nothing to do with climate.

The faces behind this Green New Deal are puppet-master billionaires, as well as Zach Exley and Becky Bond who formed the JUSTICE DEMOCRATS under the command of organizations financed by the billionaires.

The face of the Green New Deal is none other than the amazing freshman member of the House of Representatives, AOC.

And AOC now has her very own video. It's about the Green New Deal.

It's called "A Message From the Future", and is what's called "white board" video, which means the type of video in which a hand draws each scene. To much of that can be sickening, and indeed for that reason and others, the AOC video tanks very quickly.

It takes us for a ride in a row boat filled with holes.

"A Message From the Future" is 7 minutes and 35 seconds of drawings, with narration by AOC herself, about how wonderful the future will be if the Green New Deal is instituted.

While it's torture to watch the entire 7 minutes 35 seconds, try to stand the first two minutes so you get a rough idea of what this is about.

Here is the AOC video:

https://www.youtube.com/watch?v=d9uTH0iprVQ&t=1s

What is amazing about this seven minute video is its huge production cost and the mass size of the production team. As we have said many times in this exciting book, money is what makes the Global Warming fraud go round.

There were two professional directors, Kim Boekbinder, and Jim Batt. And in addition the producer, Naomi Klein, and script writer, Naomi's husband, Avi Lewis. The artist who drew the pages is none other than leading white board artist for the Left in the US, Molly Crabapple.

The video begins with a cartoon-like AOC sitting on a bullet train in the future, recounting how wonderful the world has become under the Green New Deal.

Based on the number of people working on the project, and their status, the budget for this video may have been over $ 100,000.

Soooo, my five year old daughter, who found AOC's video to be quite weird, decided to make her own video. Budget zero. I helped her a little because I studied video production with Dr. Tero Fuji at university.

And this new video or ours, in the fight against Climate Tyranny, is a free gift to those of you who now are reading this wonderful book.

Three minutes long. In which she uses, with my permission of course, my daily speeches she hears around the house about how the Green New Deal is just a power grab. It's about money and power, and not about climate.

What's really exciting is that AOC's video begins on a bullet train and so does ours...

Now without further ado... here is the link to our reply video... which is your free gift:

"THE WORLD ACCORDING TO AOC". On YouTube:

https://www.youtube.com/watch?v=cHS-B0UWJ1I&feature=youtu.be

I have to admit something. That in some ways, I have a hard time writing anything critical of AOC. She is so funny, what would we do without her in congress? Let's just briefly take a look at her latest adventure.

When returning from a two week congressional vacation, she ran up to the one square yard plastic box stuffed with commercial indoor plant soil, on the roof of her apartment. She ran up there to see if the seeds her congressional interns had planted, had produced plants. By golly they had, and this is what she reportedly said, really:

"I am SHOOK, like honestly, gardening, FOOD-- that comes out of dirt... it's Like magic " ~ @AOC

What would we do without her? Even my five year old laughed so hard at this she almost fell over.

And dear readers, AOC is the person we are supposed to put in charge of American agriculture, as part of the Green New Deal? YIKES.

Now watch a little of AOC's video first so you can see the craziness against which we are pushing back.

Then take a look at our parody video.

Our push-back video is here:

https://www.youtube.com/watch?v=cHS-B0UWJ1I&feature=youtu.be

it's part of the national artillery barrage against the Global Warming fraud.

A little something to help America wake up and get ready for the coming Ice Age 2025.

CHAPTER 19

THE GLOBAL WARMING & CLIMATE CHANGE CAUSED BY HUMAN ACTIVITY FRAUDS HAVE ALREADY STARTED KILLING US

While AOC can act in a funny way, the actual results of the Global Warming fraud are catastrophic. And we don't have to wait until 2025 to get the awful results from the lies of the Global Warming CO2 fraud.

Very directly, we've already started paying the price in lives for the false religion of Global Warming & Climate Change.

It always happens. When false gods are worshiped there is always whiplash.

In California, November 2018, in the massive forest fires, over 80 people died, probably many more... many were never found.

And hundreds of thousands of animals died. Countless numbers. And a million beautiful trees burned to the ground. All this because of the fake Global Warming & Climate Change "environmentalism".

After the fires began, the President correctly stated that a major cause of the severity of the fires was the lack of forest management.

Immediately the idiot talking heads in the Democrat Media Complex began to attack him and try to humiliate him. Fake "experts" were brought in to attack the president's correct idea that lack of forest management had led to the massive scope of the fires.

Typical was an Associated Press article, which stated that "Fire Scientists" (which could mean a fireman or even a person with a barbecue in their back yard) stated that the fires were not because of lack of forest management. But because of Global Warming.

The rule of the Democrat Party is: never let a good disaster go to waste. They always use them for political purposes to increase their power. No matter what the present and future disastrous side effects of the Democrat actions are. If a policy can help them grab power, they will use it... no matter what the cost to others.

So of course, the first move during and after the forest fires was a Democrat Media Complex attack on... the president.

However after the week of Democrat Media lies that lack of forest management did not effect the fires, soon actual forest experts explained that yes, indeed, failure to remove dead trees and massive amounts of accumulated dead brush and huge amounts of growing brush were a main factor of the fire.

And that with good forest management, these sources of fuel for the massive fires would have been removed.

The radical Democrat, Justice Democrats trained and funded environmentalists of California believe it's best "to let nature take its course". Not to cut down trees, not to remove dead trees and plants, bushes and so on.

When the electric company of California, Pacific Gas and Electric, requested permission multiple times to cut down trees near power lines, it was refused by the government of California and by the environmental radicals. This act to avoid catastrophic fires wasn't woke enough. In Global Warming mob minds, nature has to take its course. When it comes to fires anyway.

This resulted in trees and brush growing right next to power lines. And dying next to power lines.

Many dead trees and dead brush was there. Making a nice source of fuel for a fire around almost every single power line pole. It was a disaster waiting to happen. And it did.

And as living trees grew taller they actually touched the electrical components at the top. That and wind also cause sparks from such electric power line poles and steel structures.

So when the electrical sparks happened, and they always do, the trees and brush growing around power poles and steel structures, provided ready fuel for a fire from the sparks. Particularly the dead trees and dried dead brush that has not been cleared out due to lack of forest management.

President Trump suggested that "forest rakes" should have been used. And he was ridiculed by the Democrat Media Complex. Because in their absolute ignorance about nature, they though this meant sending out people with garden rakes.

It didn't. The president meant this:

The policies of the irrational lunatic California environmentalists, no cutting trees, no clearing brush, are inexplicable. And they cause destruction.

And here is a bit of news for the environmentalist religion members who are reading this book: you claim to love the trees. You even hug the trees.

But your policies are not only leading to them being burned to a crisp in forest fires in California today, but also that they will freeze out of existence tomorrow. That's what happens in Ice Ages.

As mentioned above, in California, the public electric utility, Pacific Gas & Electric, asked the state government and the environmental organizations for permission to clear trees out from around electrical wire utility poles. Not clearing out the dry bush and dead trees from around electric pole wiring would bring catastrophe. PG&E made repeated attempts to get permission to clear out the areas around electrical poles. But the lunatic environmental lobby and their representatives in California state government, would not permit it.

Thus these smart requests for forest management were refused. Let nature alone said the environmental wackos. So when some sparks flew from the electric wires, perhaps in a windy time, the dry brush and dead trees were right there to make a burning hell out of the living trees, animals, and birds, and people too.

So that's environmentalists killing trees today. They are the tree murders.

Global Warming does not exist. It is not killing trees. The environmentalists and their billionaire funders are doing it.

As for the future... failure by America to get ready for a mini-Ice Age, like the one in Europe and North America from 1300 to 1800, for for a much worse major Ice Age, will mean that many of the trees will die. If it's a major ice age, then all of the trees will die sooner or later.

As a cold weather expert and researcher who has lived in the coldest inhabited areas on earth for many years, I've seen this phenomena relating to trees and very cold temperatures.

It's part of the natural landscape, how in Scandinavia as you travel further and further north, the tree types change. The forests thin out. And then when you are at the very far north such as in northern Scandinavia, there are basically no trees at all. In winter, it looks like an off-white desert under a gray sky.

The Global Warming / Climate Change fraud mob, not content just to kill the trees, joined in during the fire to make it political. To score political points instead of solving the problem. This includes of course the Global Warming / Climate Change scientists who manage to produce completely wrong information without end.

The dean of Environmental Studies at the University of Michigan, Jonathan Overpeck, wrote that "Global Warming" had baked the forests and that Global Warming had thus caused the fires.

Sure. So instead of clearing out the 129 million dead trees and billions of dry brush clumps, let's focus on lowering car emissions to reduce CO_2.

Brilliant Dean Overpeck!

Experts who really know what happened in California said something different.

They wrote "And it's more than trees. Dead shrubs around the bottom of trees provide what is called 'ladder fuel,' offering a path for fire to climb from the ground to the treetops and intensifying the conflagration by a factor of 10 to 100", said Kevin Ryan, a fire consultant and former fire scientist at the U.S. Forest Service. In other words, lack of forest management created a tinderbox.

It's true there had been a drought, a lack of water in California for a few years. While the state was under Democrat management. Strange, isn't it, that there is no water shortage in the United States where environmental radicals are not in charge?

Water management techniques in California are more or less deteriorated and counterproductive. "Let nature take its course!", is their policy. So while it's true rainfall has been less than usual, the severe results of the drought itself was brought about by catastrophic water management techniques.

Techniques based on environmental hysteria and funding from the Left wing Democrat billionaires.

Instead of saving water through new infrastructure and policies, instead of saving the forests by forest management, the state of California environment radicals left things to go natural... in other words, to rot.

The catastrophic fires resulted, because California concentrates on efforts to reduce CO_2 in the air. To supposedly fight "Global Warming". Rather than to rationally manage their ground, plant life, and water.

While using their resources to fight a non-existent Global Warming brainwashed radical environmentalists watched the state burning around them.

There is no pleasure in saying it, but the blood of those people who were victims of the fire, of the millions of animals who died terrible deaths, and the burning of millions and millions of trees... is on the hands of the environmentalists radicals, on the hands of the Global Warming false religion movement.

The air pollution levels in California during for a long while after the fire, were the worst anywhere in the world. That's what happens when irrational brainwashed environmentalists are in control. They destroy everything. While claiming their policies would improve air quality, they ended up creating a catastrophe.

Of course pollution needs to be controlled and reduced as much as possible. It's obvious. But in rational productive ways. Here is what the madness of the California wacko environments produced.

People suffering in California, innocent people as they are who are under the control of Democrats, can thank the Global Warming / Climate Change movement movement for the worst air pollution on earth for a while.

Instead of protecting people of the state, the Democrat Party state government, and the Democrat local governments, and the radical billionaire-funded environmental movements used time and resources to combat a problem that does not exist.

To fight against an imaginary spook: Global Warming / Climate Change. And in misdirecting resources, and in letting nature take its course, they have caused an environmental catastrophe in their own state.

Struggling to show why the fires occurred, Global Warming radicals (and all Global Warming / Climate Change people are off-balance radicals, aren't they) stated crazily, that the reason was for the fires was not related to "forest management", because brush areas are not forest!

Playing with words to protect their lies. So while the state burns, they are arguing about their little white lie words. That because "brush" is not a forest, forest management would have been irrelevant. How crazy is that.

Of course brush is cleared in "forest management".

This type of misdirection, and word game, is typical for the Global Warming leadership and sadly of their brainwashed mob also.

Isn't it obvious that "forest management" which is intelligent and directed to helping forests and people and animals avoid massive fires also included managing the brush areas in and around forests? Of course.

And even though the fire area was a dry area, forest management would have cleared out the dry, dead trees and brush. And limited the impact of the fire and in fact made it controllable. Forest rakes could have helped avoid a major catastrophe.

A world leading expert on forest fires is Dr. Tim Ball. He's now a private consultant on environmental matters and was a climatology professor at the University of Winnipeg in Manitoba, Canada.

He is also at this time, the director of the Ottawa, Canada headquartered International Climate Science Coalition. Definitely one of the engineering and scientific leaders in his fields of work.

Here is what he said about the California forest fires of 2018:

"The natural cycle of forest fires creates what are called crown fires. They move through quickly, burning off dead debris but leaving most of the plants still alive.

"When governments decided to stop forest fires, they upset the natural dynamics completely. The bureaucracies, now populated by graduates of the biased environmental education system, willingly allowed the environmental extremists' demands to end the former sensible practice of cleaning the undergrowth.

Activists complained that such forest tending was not 'natural,' when it was, in fact, a reasonable facsimile of 'nature'.

So, the debris built up, leaving the forest a tinder box all ready to ignite. Making matters worse, when a fire takes hold, it now often creates what is referred to as a base fire. These fires are very difficult to extinguish—the heat allows such fires to burn into the ground and, days after a fire is supposedly out, it will flare up again."

https://townhall.com/columnists/timball/2018/12/07/extreme-wildfires-caused-by-extreme-stupidity-not-global-warming-n2537168

Who pays for the Global Warming / Climate Change craziness? It's us. We the People. Ordinary Americans who are pushed around by an insane Democrat Party, Leftist environmental movement of scientific fraud and financial fraud.

We pay the price, with our very lives. The animals of the forest pay the price. The trees pay the price.

Global Warming, the false religion, has now created a very unpleasant situation for We the People.

They have managed, by diverting resources which could have been used to get reading for the coming Ice Age, to hit us with double disasters: fire and ice.

Fire from lack of forest management and ice from the coming ice age without any preparation for it.

Behind the scenes, the Global Warming some of the politicos in California realized they have been wrong. Quietly they are trying to reverse the disaster of their Global Warming religion focused environmental policy. They recognized that forest management is needed, and urgently. The LA Times reported:

California Democrat Gov. Jerry Brown partnered with state lawmakers to introduce changes to the state's policies. The bill would grant $1 billion toward forest thinning and ease regulations on cutting trees on private property.

Now after the fire, forest management fever now sweeps California !

But that slight breeze of hope was dashed when Gavin Newsome, a dyed-it-the-wool Global Warming believer, became governor in the November 2018 election.

Dear readers, I know that you remember, in the beginning of this book, I mentioned I would try to make it a lighthearted little book.

But the coming Ice Age, and the damage caused by the false religion of Global Warming has damped that approach down a little bit.

So I'd like to make up for it, with a nice (modified a tiny

bit) uplifting and right-on cartoon by the wonderful cartoonist Henry Payne.

What the sad California forest fires demonstrate is what happens when the false path of Global Warming is followed.

The politicians, the educators at universities, colleges and schools, the protesters in the streets and government buildings harassing people, the unhinged Leftist billionaires.... all of them share a guilt.

But we can make up our own minds. Look into the matter and see the truth.

"Warming fears are the "worst scientific scandal in the history…When people come to know what the truth is, they will feel deceived by science and scientists."

United Nations Intergovernmental Panel on Climate Change, Japanese Scientist Dr. Kiminori Itoh, an award-winning PhD environmental physical chemist.

The way out is escape from the false religion of Global Warming & that Climate Change is caused by human activity.

"When we're prevented from managing our forests by these radical environmentalists — they've had lawsuit after lawsuit, they have somehow promulgated to let nature take its course — this is the consequence of letting nature take its course,"

"We need to go back to prescribed burns late in the season so you don't have these catastrophic burns, remove the dead and dying timber, sustainable harvests, get the small mom and pop mills back where they're grazing the forest and return to healthy forests,"

"You look at Finland. I had an opportunity to live in Germany. Germany has the Black Forest — their forests are healthy, they don't have the catastrophic burns because they manage the forests."

"And I will lay this on the foot of those environmental radicals that have prevented us from managing the forests for years. And you know what? This (disaster) is on them."

Words of truth by Ryan Zinke, former Secretary of the Interior, United States of America.

Just a note: Ryan Zinke was forced from office as Secretary of the Interior by a flood of lies against him. By a flood of false ethics complaints from various Left wing funded organizations. Environmentalist wackos.

This is a traditional rotten tactic by the political Left mob. They lose elections then try and force out the winners they don't like with false ethics complaints against them.

The victims are forced, according to US and state statutes, to defend themselves with their own time and money.

This method was utilized to destroy, immobilize, and terminate the excellent Alaskan governor's career of Sarah Palin. And also for example, the proven fraudulent accusations by prosecutor Andrew Weissmann in jihad against Senator Ted Stevens, also of Alaska.

Ted Stevens was finally found innocent, and it was shown that false evidence was used against him and that Andrew Weissmann had withheld exculpatory evidence from the courts.

This is how the left wing environmental people work.

Dear readers of this book, most of you have no connection to the political Left and hysterical environmental Left, but for those readers who are Left-leaning, get out of there while you still have time. The Democrat, Justice Democrat environmentalists are rotten to the core. And if you don't leave them, you will become rotten too.

For those of us who are conservatives, constitutionalists, we who support President Trump, we have to start pushing back and crush these efforts by Democrats to destroy those who are telling the truth. This politics of destruction and smearing by the Democrat Party is typical in their Global Warming actions.

They are pushing the country into national suicide, and this book is a guide to defeating them politically, judicially, and in the media.

The United States and its people are not ready for the coming Ice Age. Not ready for 2025.

The barrier, the blockade against preparedness is the Democrat Party and its mob.

CHAPTER 20

WHAT ABOUT THE GLOBAL WARMING BILLIONAIRES?

In mid-November 2018, 200 young protesters descended on the office of Nancy Pelosi in Washington.

About 40 of them sat on the floor nicely. A nice sit-in.

It's a good thing they did it at the beginning of November, because it became so cold at the end of the month that their poor little behinds would have frozen to those stone floors of the capitol building.

Anyway, the Global Warming sit-in was on a workday. But they were in Nancy's office anyway. So why weren't they at work? Well, in a sense they were.

Global Warming protests, as with many other Left wing protests are bought and paid for by Left wing billionaires and multi-millionaires.

Yes, the dreaded 1%, that the protesters claim to hate, is financing the fraudulent Global Warming movement. Without the money of the 1% billionaires and multi-millionaires on the Left, not much would be happening.

In this chapter we are going to focus on the billionaires and millionaires WHO MADE THEIR MONEY from the Global Warming hoax.

Number one of course is the so-called father of the Global Warming monster, Al Gore.

Al was living on his pension. Not a bad one because he held some high paying jobs, including Vice President under Bill Clinton. But it wasn't really enough to support his lifestyle, and he felt awful after losing the presidential election to George Bush. And so he turned to Global Warming as a way to start the big bucks flowing in.

Al Gore made his first big money from the Global Warming fraud, a little donation from George Soros funded organizations.

About 2004, Soros ordered Gore to open the way and spread the word for a Global Warming movement. Many assume the purpose of this movement was to spread instability and overthrow of existing governments.

The little donation amounted to $30 million.

So immediately Gore went from driving an old Chevrolet family coupe

to a new electric car called the Chevrolet "Spark".

Moving up in the world must have tasted good to Gore. And he then went after more and more money on the Global Warming scam track.

The first money after the donation, was brought into the first and longest lasting major financial Global Warming scam, with a science fiction movie about... you guessed it, Global Warming.

As hype for the movie, he declared in 2006, that in ten years the world would have reached a point of no return. That in 2016, we would all become toasted marshmallows as the result of Global Warming unless radical steps were taken without delay.

The Leftists of Hollywood foamed from their mouths with glee, and managed to wrangle an academy award for Gore's science fiction film, which was falsely billed as a "documentary".

Not only that, Frostbite Al managed to rake in about 1 million dollars from a Nobel Prize. A prize that he got for a "documentary" film which was entirely fiction.

And guess what, after all the hype about the end of the world in 2016 by Gore, when the ten year doomsday clock ticked the last seconds, the world was still here.

And no Global Warming at all had taken place.

No boiling rivers and lakes and oceans.

No New York flooded by rising sea levels. Everything was OK.

So are we finally finished with the crazy predictions by Global Warming hysterics?

Here's what David French of the National Review journal wrote:

"Apparently not. Being a climate hysteric means never having to say you're sorry. Simply change the cataclysm — Overpopulation! No, global cooling! No, global warming! No, climate change! — push the apocalypse back just a few more years, and you're in business, big business.

Being a climate hysteric means never having to say you're sorry."

Gore just happened to invest in companies that benefited from his movie. Solar companies, wind companies, and so on. Later many went bankrupt, but for him, in it for the fast buck, it made him a multi-millionaire.

But while that might seem OK to those who worship his fiction, how he made his big cash-in, came from a deal he made later with the Saudis.

Yes those same Saudis who produce all those carbon fuels: oil and gas. He took their money happily. And it's described on the next page.

In addition to Al Gore, another man made rich by the Global Warming hoax is Rajendra Pachauri.

A sort of fake scientist who was head of the United Nations fake panel on Climate Change. He became a multi-millionaire from consulting fees, his work at the US, books, tours, investments into stocks that were related to alternative energy, like windmills.

As the leading "scientist" at the UN for climate change, Pachauri said 2007 we only had four years to save the world.

His "science" was as a railroad engineer. And based on that he said:

"If there's no action before 2012, that's too late."

"What we do in the next two to three years will determine our future. This is the defining moment."

He was the hero of the Global Warming movement, a guru to the touchy-feely brand of supporters. He was from India and looked the guru role.

But he got a little too touchy-feely with some women he worked with, and they didn't like it. He and had to slink out of the UN job, and drop off the radar screen. Sexual misconduct, you know.

Al Gore also decided to get out of the Global Warming game with piles of cash, when he sold his Global Warming assets, mainly a cable TV network that no one watched, to the Saudis.

The Saudis wanted a way to get their brand a more friendly reception in the US and thought that Gore's TV channel would do it. It didn't because no one actually watched it. But the Saudis paid multi-millions for it.

And what about this. Where did the Saudis get their money to pay Al Gore? From fossil fuels! From oil and gas! The very enemies Al Gore had declared war upon. Yikes.

So here dear readers, is a message to anyone who is protesting on the street or in Nancy Peolsi's office on behalf of the fake Global Warming religion....

Your protests help the 1% ! A very large group of 1% that makes its money by scamming.

CARBON CREDITS, THE HIGHWAY TO BECOMING A MILLIONAIRE... BASED ON FRAUD

Because not everyone can make a Hollywood movie, there was developed another avenue to Global Warming riches.

Thousands of millionaires have been created through this new way, the CCC... the Crazy Carbon Credits scheme.

What is "Carbon Credits"? Well it's like bitcoins in overdrive.

Many Left wing governments in North America and the EU bought into the Global Warming hoax. As a result they pointed a finger at their own industries and factories. "Bad boys!" they said. And they said, "You're belching out all that nasty CO_2. And you're now going to pay for it!"

So as a sort of punishment, laws were made that industries and factories that had CO_2 emissions had to stop doing that (which means they shut down) or to buy "Carbon Credits" from people or organizations who were not producing CO_2.

For example, if there is a forest in South America that just sits there and is not producing pollution, they get Carbon Credits, and can sell them to factories and industries.

Billions of dollars in Carbon Credits are bought and sold every year. And people can make a lot of money on them although the whole system is totally fake.

It also damages industry and jobs in many of the most industrially advanced countries, such as North America and the EU. Carbon Credits is a tax on them. They have to buy carbon credits to keep going.

Let's take a look at a few because no one could believe they exist without seeing them. And no, they are not from a Monopoly game, they really exist...

The certificates cost a lot of money to factories that are forced to buy them, and

raise the cost of manufacturing goods.

Then those companies that are forced to pay for Carbon Credits can't compete for example, against factories in China, that don't pay any carbon tax. The government of China, say what you will about it, knows Global Warming /Climate Change is a total fake.

The Global Warming / Climate Change movement is financed by the 1%. Billionaires. The people what the protesting mob claims to hate. It's really crazy, isn't it.

CHAPTER 21
BIG GLOBAL WARMING MONEY IS VERY REAL!
IT'S A HUGE AMOUNT...
AND WE AS TAXPAYERS ARE PAYING FOR MOST OF IT

Sometimes when we want to find the reason for a sneaky political movement for telling lies, good advice is to "follow the money".

Now regarding Global Warming / Climate Change finance, it's the biggest financial scam in history.

Hundreds of billions of dollars, yes billions, have flowed into the pockets of promoters.

How?

THE LAWYERS

First of all, lawyers are raking in money by the truckload.

They, for example, sue big oil companies such as Exxon for destroying the planet. Sue them for not following the Global Warming commands, and for not whipping themselves to death in public.

The funny thing about those lawyers who can each make millions and millions of dollars on a single case, is that they are sort of creepy about the money when they manage to win in court or by settlement out of court.

Usually more than one law firm is involved in these very large-scale cases, such as against Exxon and other energy companies.

And when the lawyers get paid, sometimes only one firm gets paid and they are supposed to pass on payments to the second law firm. But they don't.

Then the Global Warming lawyers end up suing each other.

That's right. The Global Warming key law firms sue each other almost as much as they sue energy companies. Sometimes the fight over the money among the lawyers goes all the way to the supreme court.

MORE ON CARBON CREDITS

But the tens and hundreds of millions of dollars to lawyers are just peanuts compared to the center of the money scam which we briefly looked at in the preceding chapter: CARBON CREDITS

Here is what they are, in the plain language that we use in the PRIMER ABOUT GLOBAL WARMING BOOK.

They are pieces of paper. And that paper isn't money but it can be worth huge amounts. What happens is that some governments, such as the European Union, believe in Global Warming, and so punish the industries in their own countries.

If a factory has manufacturing processes which create CO2, they have to find a way to offset that production of CO2. There are only two choices under the Carbon Credit tyranny.

Either the factory can shut down, or they can buy Carbon Credits from someone who is not producing CO2.

So he factory stays in business by buying a CARBON CREDIT from some place or organization or company that takes up geographic space, but does not produce CO2.

For example a steel company that produces a lot of CO2 can buy carbon credits from a forest somewhere that is just sitting there. Or from a company that produces something else and has less than the allowed CO2 outputs.

So the forest owners and the company producing candy, producing less CO2 than they are allowed, sell their CARBON CREDITS to companies which are supposedly polluting by going over the legally allowed CO2 limits.

This of course hurts industry in developed countries that agree to this useless system. Useless because as we have seen in this fun and wonderful book, that CO2 has no effect on raising or lowering temperatures.

But billions and billions of dollars every year are traded on licensed CARBON CREDITS exchanges in many countries. The people running these exchanges and working in them make billions, and the sellers of CARBON CREDITS make billions as well.

It's really a way to transfer wealth from high tech industrialized countries to countries that have production and industry dating from 150 years ago.

Carbon Credits don't help the environment, or effect global temperatures one bit. But because it's a way to make a lot of money, a new type animal on the earth created by "Global Warming" is called the Money Grubber.

And this new animal, Money Grubbers, love Global Warming as a money machine to keep their yachts and Farraris humming.

That's right! Global Warming Money grubbers.

So we see, that all the lawyers, and all the CARBON CREDIT people, and the ones pushing cheap wind turbines and solar panels that don't work very well, who are making billions of dollars, view Global Warming as simply a money machine.

GLOBAL WARMING TAXES, ON WE THE PEOPLE

And guess what, it's not only factories in a country that need to buy Carbon Credits.

For those countries that are part of international Global Warming agreements such as the Paris Accords (which thank goodness President Trump got the USA out of), the people of that country have to in a way buy carbon credits. Every man, woman, and child. And they pay it in the form of new taxes.

A country, such as France, which is the poster child for dupes that bought into the Global Warming fraud... for example, produces more CO_2 than it saves, so every family has to pay a tax.

And that tax is called, guess what, a CARBON TAX.

Who actually does the paying? The unfortunate working people have to pay it. They pay taxes and this tax money on their country's CO2 surplus. That money goes to their government, for example in France, and it disappears there forever.

And guess what, people don't like paying this tax.

Usually Carbon Tax is a tax on fuel. Fossil fuels such as oil, gas, coal. Fuel that they use to heat their apartments and homes, fuel that they use to drive their cars, to generate electricity, and fuel that industry needs.

The idea of these insane taxes is to make fossil fuels so expensive that no one will use them.

People will have to stop driving their cars, heating their homes, using electricity. So it makes people miserable and pushes them into the hell of poverty.

It hurts people also because it takes their hard earned money, and has zero good results.

Gasoline for cars for example, can almost double in price as a result of the Carbon Tax.

People are so upset about it in Europe that they light their cities on fire to stop the CARBON TAXES.

This is hidden from the US public by the Democrat Media Complex. TV, newspapers, websites that tow the tyrannical line of lies about Global Warming for the Democrat Party.

Ordinary people realize that the carbon taxes are totally crazy, hurt them and don't help them, and certainly don't have any effect on the temperature of the planet.

Prime Minister Macron of France is a big fan of Global Warming. So he committed his people to paying a high CARBON CREDITS TAX.

Here are some photos of the response of ordinary people to the crushing Global Warming tax on their gasoline, their home heating costs, on electricity... dumped on their head by their leftist

132

ideology government. A tax on something that does not exist... Global Warming & Climate Change Caused by Human Activity. It's only Halloween.

Welcome dear readers to beautiful central Paris! See the beautiful sights. Burning tires, tear gas... During the continuing carbon tax riots and protests... off and on again for months.

Americans who have been tricked into supporting the Global Warming false religion should understand that their support is going to cost them something. Or they will wake up one day and find themselves in the situation of the people of France. Carbon Taxes on everything they need to live. So much taxation, they can't live.

If, goodness forbid it, that the Democrat Party should ever gain full control of the government of the United States, you can expect your fuel, electric and other costs to go up so much, you can not afford to live.

For those Leftists reading this book, or having the book explained to them simply by a Republican / conservative friend or family member, that means you can't put gas in your car. You can't turn on the light or cook anything. Or heat your home.

Higher cost bus tickets. Higher cost to charge your mobile phone and Ipads.

In short, the cost of every product you use (not just the basics) will go up also, because energy is needed by factories to produce products.

People will buy less products because they do not have to money to afford their previous purchases they had before the CARBON TAX. So factories will have either to close or lay off workers. And many many people will lose their jobs. You could lose your job as well.

Global Warming taxes are going to cost you. Look at Europe and see how the people there are reacting to the Global Warming insanity of their governments.

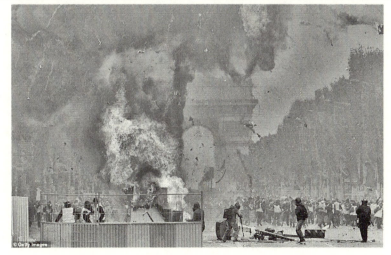

133

Photo Above: Welcome to the beautiful Champs Elysee and Arch of Triumph in Paris! With a little added ambiance from Global Warming Tax protests

PUTTING TAXPAYER MONEY INTO FAKE GLOBAL WARMING ALTERNATIVE ENERGY COMPANIES, ANOTHER FINANCIAL SCAM

There is a funny little Global Warming / Climate Change scam. New companies connected to Democrats jump into alternative energy. And although their products and technology sometimes look nice, they are junk. And they don't work, but still they are getting a fortune in taxpayer dollars.

For example President Obama approved a $500 million unsecured loan to a new solar photo voltaic company called Solyndra.

Getting government money is easy to get, because your fake company supposedly will help solve the Global Warming problem when Democrats are in control.

The company, Solyndra, which received **$500 million** in our taxpayer money, went bankrupt. And that $500 million which belongs to We the People disappeared down the Obama government fake alternative energy toilet.

That company which received $500 million of We the People's money never produced energy on more than one or two rooftops or small fields. And collapsed because of lousy technology and even worse business practices.

We usually refer to such things as a scam. And that's what Global Warming Money is. But there is a lot of it, so it attracts a small army of money grubbers who will do anything for a chance to get it.

Some people say that includes Al Frostbite Gore.

And crooked companies with connections to the Democrat Party are not the only ones receiving truckloads of taxpayers' money. The Global Warming fake scientists we have discussed in details in this book, are also waiting in line for the United States money trucks to show up at their homes.

Endless truckloads of taxpayer money have gone, and are still going to so-called "scientists" and "researchers" who support Global Warming with their fake studies.

From UN researchers, to university professors, to private research companies, the money has come in like a hurricane. The fake scientists are like dogs snapping at the dollar bills as taxpayers' money flies into their open mouths.

In one way, Global Warming hysterics have been right. There has been an unusual amount of flooding.

But if hasn't been a flood of water. It's been a flood of taxpayers' money to the researchers and authors of fake research and articles and papers.

And that money is not peanuts, it's BILLIONS OF DOLLARS YEARLY.

And in another way, Global Warming is indeed a great mystery. But the mystery is financial.

How is it possible that the protesters who are active for Global Warming, and who supposedly hate the 1%, can be part of a movement that belongs to and is paid for by the 1%?

Of course they can spend their money as they wish. This is just to let the Global Warming / Climate Change protesters know who is paying for their movement. Usually through front groups they support financially:

Tom Steyer... billionaire

Jeff Bezos... billionaire

Michael Bloomberg... billionaire

George Soros...   billionaire

They are among the puppet masters. And apparently the protesters are the puppets.

Many of the protesters have their expenses (such as transportation and room and board) or even salaries for being on the street, or on the floor for sit-in, by Left wing environmental organizations. Those organizations are funded by the 1%.

Yes dear Global Warming protesters, many of those organizations who support you are bought and paid for by the 1% billionaires. Where did you think the money you are paid comes from? It's from the 1% you claim to despise.

Being a Global Warming protester is like being someone who claims to hate carbon footprints, but then drives 50 miles in their polluting car to get to a demonstration. Why not follow the advice of AOC and go by foot? Horses? Sorry .. no. They fart methane. And uuupps... for that matter... so do you.

How Global Warming protesters that hate the 1%, can be part of a movement that is run and financed by the 1% is only one of the Halloween continual mudslides of Global Warming.

WHY A SMALL GROUP OF SCIENTISTS LIE THROUGH THEIR TEETH

The small group of scientists who sound the false alarm about Global Warming are not crazy. They're greedy.

Like sharks in a tank, waiting for the keeper to dump in some more dead fish for them to eat.. they have open jaws. Waiting for the government billions yearly and the money from the Leftist billionaires. Soros, Steyer, Bezos, Bloomberg and others. Feeding the shark-like "scientists" who are in a feeding frenzy for money..

To get some idea how much money is pouring out of governments and organizations of billionaires into their gullets, just take a look at these financial figures. They're unbelievable but true:

"According to a recent report by the U.S. Government Accountability Office, "Federal funding for climate change research, technology, international assistance, and adaptation has increased from

$2.4 billion in 1993 to $11.6 billion in 2014.

With an additional $26.1 billion
for climate change programs and activities provided by the American Recovery and Reinvestment Act in 2009."

Yes, you are reading that right.
It's a total of nearly 40 billion dollars from the US government alone. Obama, what have you done with our money? Just think how many children living in poverty could have escaped it with that money.

But getting back to the Global Warming /Climate Change scientists... by biting into this money pot with their little yellow teeth, the small group of fake and lying scientists get very rich by telling lies.

Let's look at it this way. If there are no fires, there won't be any fire department.
If there are no fears about Global Warming, no one will hire these cardboard scientists to do cruddy research and papers.

So they tell lie, after lie, after lie. To stoke by lies, fear about Global Warming & Climate Change being caused by human behavior, to keep the money coming.

In return for screwing up some math and computer models, which are a fake and show Global Warming / Climate Change, the moon is made of cheese, whatever, they can get rich. Really rich. It's as simple as that.

CHAPTER 22

GLOBAL WARMING DISASTERS... DREAMS OF THE POLITICAL LEFT THAT NEVER COME TRUE

WHAT TO DO ABOUT RISING SEA LEVELS?
RELAX, IT'S A FAKE LIKE EVERYTHING ELSE

WHAT ABOUT THE GOVERNMENT GLOBAL WARMING REPORT, SPRUNG AS A TRAP ON THE PRESIDENT IN 2018, WHICH FORTELLS DISASTER?
RELAX IT'S A FAKE TOO.

Sea Levels. They try to scare us about sea levels rising. US cities are spending a fortune to lift entire neighborhoods, and to build retaining walls against a high sea level that isn't coming.

Dear readers, we can relax:

LEADING CLIMATE SCIENTIST, SCIENTIST SAYS SEA LEVEL PROJECTIONS ARE BUNK

The church of climate change isn't going to be happy with this one.

According to the Daily Caller News Foundation, noted climate scientist Judith Curry thinks that the climate change alarmists releasing doom and gloom predictions of climate catastrophes should stop the hysterics:

> "Projections of extreme, alarming impacts are very weakly justified to borderline impossible," Curry told The Daily Caller News Foundation.
>
> Curry's latest research put together for clients of her consulting company near the end of November, looks in detail at projections of sea level rise. Curry's ultimate conclusion: "Some of the worst-case scenarios strain credulity."
>
> "With regards to 21st century climate projections, we are dealing with deep uncertainty, and we should not be basing our policies based on the assumption that the climate will actually evolve as per predicted," Curry told TheDCNF.
>
> "Climate variability and change is a lot more complex than 'CO2 as control knob'," Curry said. "No one wants to hear this, or actually spend time understanding things," Curry said.

This isn't the first time an excellent and honest scientist has spoken out about alarmists putting false science and inaccurate predictions out for public consumption to cause panic.

Panic which has the intent to move forward the power grab of the Democrat Party over the nation.

Among the many brilliant scientists speaking out against error in Global Warming scientists, is Princeton University physicist William Happer who had to correct Bill Nye live on CCN about his ocean heating models...

Which Nye didn't take well and demanded networks like CNN stop putting people like Happer — an actual scientist — on to give their opinions.

Nye himself is not a scientist. He's an actor.

Curry has called the latest alarmism of the Global Warming storm troopers, the latest about the oceans rising to submerge the coastal cities, **"unlikely to impossible"**

According to the Daily Caller:

> Alarming sea level rise predictions are based on "a cascade of extremely unlikely-to-impossible events using overly simplistic models of poorly understood processes," Curry wrote in her report.
>
> Current sea level rise is well-within natural variability of the past few thousand years, according to Curry. Curry said coastal communities should base their future flood plans on likely scenarios, such as one to two feet, rather than high-end scenarios.
>
> "There is not yet any convincing evidence of a human fingerprint on global sea level rise, because of the large changes driven by natural variability," Curry wrote. "An increase in the rate of global sea level rise since 1995 is being caused by ice loss from Greenland."

Curry thinks these Global Warming / Climate Change scary stories about the future are actually agenda-driven models. And not based in reality.

https://www.redstate.com/brandon_morse/2018/12/10/noted-climate-scientist-says-global-warming-alarmists-predictions-borderline-impossible/

AND NOW FOR MORE FUN:

HERE ARE SOME DETAILS ABOUT THAT FEDERAL GOVERNMENT SCARY GLOOM AND DOOM GLOBAL WARMING REPORT OF 2018

RELAX:

IT WAS A FAKE ! IT WAS AN ATTEMPT TO BLINDSIDE, TRAP, EMBARRASS,

WEAKEN, SMEAR, AND TRICK THE PRESIDENT OF THE UNITED STATES.

IT WAS A POLITICAL STUNT, NOT A SCIENTIFIC STUDY. A DIRTY TRICK FOR WHICH HOPEFULLY THE GUILTY FRAUDSTERS WHO WROTE THE REPORT WILL BE PUNISHED.

SIMILAR TO MANY IF NOT ALL GLOBAL WARMING / CLIMATE CHANGE STUDIES FROM THE LEFT THESE DAYS, THE REPORT IS A TOTAL FAKE:

Climate expert Dr. Roger Pielke Jr. had this to say about it and how dishonest this sham report is:

"The claim of economic damage from climate change is based on a 15 degree F temp increase that is double the "most extreme value reported elsewhere in the report."

The "sole editor" of this claim in the report was an alumni of the Center for American Progress, which is also funded by Tom Steyer."

Climate analyst Paul Homewood characterized the fake report, which was written by Leftists to damage President Trump, in this way:

'Cherry picks' a few bad weather events...extrapolates using the most scary scenarios'

Climatologist Dr. Pat Michaels on the report:
'Systematically flawed' – Report 'should be shelved'

Dr. Ken Haapala:
'The global warming chorus immediately seized on the new USGCRP report claiming the Trump administration is contradicting President Trump's claims about global warming.

Amusingly, some of the chorus interviewed people who worked on the USGCRP, who were political appointees under the Obama Administration.'

Greenpeace co-founder Dr. Patrick Moore:

"The science must be addressed head-on. If POTUS has his reasons for letting this Obama-era committee continue to peddle tripe I wish he would tell us what they are."

In fact, the President was blindsided. He didn't know anything about the content of the report before it was released and the press ate it for dinner.

Dr. John Dunn:
"Two years into the Trump administration it is sad to see this 400-page pile of crap."

Climate Depot's Mark Morano:
"It is a political report masquerading as science. The media is hyping a rehash of frightening climate change claims by Obama administration holdover activist government scientists. The new report is once again pre-determined science."

"The National Climate Assessment report reads like a press release from environmental pressure groups — because it is!

Two key authors are longtime Union of Concerned Scientist activists (Left-wing lying hacks), **Donald Weubbles and Katharine Hayhoe."**

The new book, The Politically Incorrect Guide to Climate Change, MIT climate scientist Richard Lindzen wrote of the National Academy of Sciences:

"Regardless of evidence, the answer is predetermined. If the government wants carbon control, that is the answer that the Academies will provide."

Scientists have ripped the new federal climate report that was published in an attempt to embarrass and smear President Trump, as:

"tripe"
"embarrassing"
"systematically flawed"
" the key claim based on study funded by
  Steyer & Bloomberg"

And finally dear reader, to help us all relax as the increasing flood of Global Warming lies flows down into the sewer, Astrophysicist Dr. Willie Soon, unloads on the Global Warming scientists intentional false math.

Dr. Jeffrey Foss wrote about it in a comment entitled, "Dr. Willie Soon versus the climate Apocalypse"

"What can I do to correct these crazy, super wrong errors?" Willie Soon asked plaintively in a recent e-chat. "What errors, Willie?" I asked.

"Errors in Total Solar Irradiance," he replied. "The Intergovernmental Panel on Climate Change keeps using the wrong numbers! It's making me feel sick to keep seeing this error. I keep telling them – but they keep ignoring their mistake."

Astrophysicist Dr. Willie Soon really does get sick when he sees scientists veering off their mission: to discover the truth. I've seen his face flush with shock and shame for science when scientists cherry-pick data.

http://www.cfact.org/2018/12/02/dr-willie-soon-versus-the-climate-apocalypse/

In short dear readers, as we have said from the beginning of this new fun book about Global Warming, the science is junk.

Crap bought and paid for by Global Warming billionaires, by our very own previous Obama government, and by the hold-over climate scum still left in the new Trump administration from before.

Once we ferret out the Obama hold overs scum, things will improve. And the present administration is working on this now.

So, just relax. All the so-called scientific Global Warming / Climate Change hysteria is nothing more than a pack of lies dressed up in reverse Halloween costumes. To make monsters into humans.

CHAPTER 23

WHAT CAN WE DO? IF ANYTHING, TO PREPARE AMERICA FOR THE COMING ICE AGE?

Because we are entering an Ice Age, not a warm or hot age, it is unfortunately possible that as many as half the population could die of exposure to cold.

There will be little or no heat in homes of course in major cities such as New York and even in California. Cities which are the vipers' nest of the Global Warming Fraud leaders..

And We the People about who the Democrat Party, Justice Democrats lie to, about caring so much about the climate, will die from the consequences of the ultra freezing temperatures have on water and water delivery infrastructure.

Water delivery systems will stop working when the pipes freeze shut or burst. There won't be any bottled water or Starbucks coffee either.

Elevators won't work in an ice age. Electrical generating power stations won't work, cars and trucks won't work, and on and on. Cities will not be able to clear their roads and highways of snow and ice. Everything stops. AOC and the rest of the Democrat Party, Justice Democrat / Democratic Socialists that are pushing the great fraud, will be locked down. They won't be worried about cow farts at that time. But how to get drinking water or how to get warm.

That the so-called Global Warming movement represents a significant threat to the United States is clear.

Because it is stopping the country from getting ready for what is really happening.

What is around the corner, between 2019 and 2035 as the start date.

It's like a tsunami is coming and the Democrat Party, Justice Socialists / Democratic Socialists are yelling:
"Go to the beach and enjoy yourselves!"

Bad advice.

The Democrat Party, Justice Democrats / Democratic Socialists must for sake of political correctness, change their name to the Lemming Party.

Lemmings are cute little animals that commit mass suicide.

Without preparation for brutally cold weather, **the best we can hope for is societal collapse, starvation, and violence**. The 500 year Mini-Ice Age in Europe and North America from 1300 to 1800 was like that.

Or alternatively, if the ice age is a brutal one, a full ice age, we will be wiped out. Or rather, frozen out over a long period of time..

Resources which could help us get ready for increasing freezing temperatures, are being squandered on a problem that does not exist. The fake Global Warming due to human activity, is eating up time and money we need to get ready for the Ice Age.

Guess what? The new ice age could start in 2019 or 2020. And certainly before 2035 based on the sun's thermal output.

The sun is already moving clearly into a solar minimum. Sun spot are disappearing, ocean currents are at their lowest flow rate in 1,500 years, and the earth is radiating its warmth into space at an alarming rate.

We have had some inkling what the start of a Little Ice Age it will be like already. Welcome to sunny, tropical El Paso, Texas in early November 2018.

Yikes, it looks like Alaska. Doesn't it?

But even that was nothing compared to Thanksgiving of 2018 for much of the US, where temperatures were 25 or even 30 degrees colder than usual.

Some cities may saw their coldest Thanksgiving Day on record.

Lows were 15 to 30 degrees below average Thanksgiving and Black Friday. Shoppers hands froze to their credit cards.

And it wasn't just a freak storm that occurred for Thanksgiving. It's a taste of what we are going to get regularly after we go over the cliff into the valley of the Ice Age. Estimated 2025, but at latest 2035. And it may come before that. It's at the door.

Here is what it tasted like:

CBS NEWS CHICAGO:

One of the worst November snowstorms in Chicago history hammered the city and suburbs Sunday night and early Monday, dumping more than 7 inches in the city, and up to a foot in some northwest suburbs.

With winds that gusted up to 50 mph at times, the storm knocked down trees and power lines across the Chicago area, leaving hundreds of thousands of homes and businesses without power. The storm also forced airlines to cancel hundreds of flights on the busiest travel day of the Thanksgiving holiday weekend.

According to ConEd (the power company in Illinois), approximately 339,000 customers lost power during the storm. That means 339,000 houses or apartments. Let's assume an average of just 3 people per customer, that's one million people without power.

People across the city were praying that Global Warming would finally start. But the false religion prayers failed. Freezing continues.

So what can we do now ?

First, vote Republican, like your life depends on it. *Because it does.*

Second, brave people have to work to politically, judicially and through the media, crush the illegal voting frauds committed by the Democrat Party, Justice Democrats / Democratic Socialists.

Illegal voting is a serious crime and carries some serious jail time. So why not work with private volunteer Republican attorneys, with your local and state Republican Party and local law enforcement, with district attorneys and US government attorneys, in cooperation with the Republican party, to stop and to crush the illegal voting machine of the Democrat Party.

We saw Democrat election fraud in action in Florida, Georgia, and California clearly in the mid-term elections of November 6, 2018. But that is just the tip of the iceberg.

University studies estimate that around two million eight hundred thousand illegal votes are cast in elections nationwide for presidential and mid-term elections. And of course for state-wide and local elections also. Yes 2,800,000 illegal Democrat, Justice Democrat votes.

The number of illegal alien voters is increasing like a firestorm. In 2018 an additional 1.5 million illegal immigrants entered the US. Mainly through the open southwestern border that the President is trying to close.

The firestorm of illegal voting and other Democrat Party election fraud has to be extinguished, and crushed into cold ashes.

Third, work with the Republican party, elected officials, local and state government officials, to get ready for the coming freeze.

And start programs to let it be known throughout the land, far and wide, that those who advocate for programs to fight Global Warming / Climate Change are actually driving the country into a frozen suicide.

For those of you on the political Left, all those nice illegal aliens that the Democratic Party is bringing into the country are not used being in to the very cold weather that is coming.

So Democrat, Justice Democrats / Democratic Socialists... readers of this book: **Save The Illegal Aliens!** Save them by getting ready for the coming ice age.

Perhaps the Democrat, Justice Democrats / Democratic Socialist people will warm the illegal alien guests across our border, up by opening their homes to them? Don't count on it. They talk the talk, but don't walk the walk.

About action three, getting the infrastructure of the country ready for the cold, it includes among other things, hardening the water and electric / gas and water systems against long term sub-freezing temperatures.
It requires specialized systems, which for example the author of this book is

familiar with in Scandinavia. Where everything is underground a few yards to help insulate the water and power and gas from the cold. The excellent Scandinavian cold weather systems are very different than used in almost all of the US.

There is a lot of work to do to get the energy and transportation infrastructure ready for the cold.

This includes setting up a snow clearing system for roads and highways that actually works. For example in New York City, and many other cities, the existing snow removal system is basically 100 years old in technology, and ineffective in the case of heavy snow or ice.

The vast majority of US cities have no snow removal capabilities at all. But they will have to plan and put a basic system in place as soon as possible.

Key is getting solid constitutional conservative smart (like the Republican Freedom Caucus in congress) into local, state, and federal government positions of power. People who are not fooled by the Global Warming and Climate Change fraud.

People who will be ready to tackle the huge challenge of freezing temperatures when the sun's energy drops soon to the critical Little Ice Age level. That means Vote for Republican conservative constitutional candidates.

And elect those Republicans who are serious defenders of the country and its people. Not fake Republicans, not RINOs, but real constitutional, conservative, supporters of the president ones.

Support the President. He has it right about the climate.

Your future, your family's future, and country depends on it. You can do it.

And you can feel good while getting ready for the new ice age. No guilt or doubts, because CO2 is good, not bad. It's the building block of life. Plants can't live without it.

Right now the CO2 level on earth is just about 420 parts per million. In other times of beautiful climate and flourishing plant and animal life, the level has been 1,200 parts per million. Four times higher. Yet the climate was wonderful then.

So even those of you who were blinded and were radical Global Warming / Climate Change supporters, you don't need to feel guilty if CO2 levels go up while you are preparing for the new ice age.

Without Carbon Dioxide we would all die within a few months on earth.

It's necessary for our life here on earth.

Plants, which are the source of all food for life on earth, use solar energy to produce carbohydrates from carbon dioxide and water in their life process called photosynthesis.

Without CO2, vegetables, trees, all plants would die. And that's not only sad for the plants, but we wouldn't have any food, and we would all die as well.

And definitely we wouldn't have any Pepsi or Coca Cola either because they get their bubbles from CO2 being injected into them at the factory.

And without greenhouse gasses to hold heat into the earth's atmosphere that enters from the sun, we will all soon freeze to death. CO2, and all the natural greenhouse gasses are a blessing.

But the ice age is coming, despite the best efforts of CO2 to save us. The ice age is coming because the sun is growing way cooler.

And as a final word, now that we are all on the same page and understand that we are heading into an ice age.... here is some message from among the greatest climatologists in the world,

So be happy you are in love now with CO2, it's good for the planet. Have a wonderful life. Listen to what world-leading scientists and even polar bears have to say:

"**CO2 emissions make absolutely no difference one way or another....
Every scientist knows this, but it doesn't pay to say so...
Global warming, as a political vehicle, keeps Europeans in the driver's seat and developing nations walking barefoot.**"

- Dr. Takeda Kunihiko, vice-chancellor of the Institute of Science and Technology Research at Chubu University in Japan.

And:

"**Warming fears are the "worst scientific scandal in the history...When people come to know what the truth is, they will feel deceived by science and scientists."** -

UN IPCC Japanese Scientist Dr. Kiminori Itoh, an award-winning PhD environmental physical chemist.

To add to these stirring messages from two of the world's leading scientists who have spoken about the Global Warming fake, let's take and interesting turn and give the polar bears the last word.

Here is little Miki Atiqtalaaq, to give a message to all of us:

"Dear Human beings. We don't want you to freeze to death.

So please, please ge*t* ready for the coming new ice age. And do all you can to keep it from getting too cold on earth.

Because if gets too cold then all the polar bears will die too!

Save the planet from the coming ice age!

Please! We polar bears sense it is coming, and soon. Let's make it through together !

   Hopefully,
   Miki Atiqtalaaq Little Polar Bear

CHAPTER 24

**SAD NEWS FOR GLOBAL WARMING RELIGIONISTS.
AND FOR THE REST OF US ALSO.**

THAT THE NEW ICE AGE IS STARTING SOMETIME BETWEEN 2019 AND 2035. VERY LIKELY ABOUT 2025

SEVERAL MEASUREMENTS, THE OCEAN CURRENTS, SOLAR ACTIVITY, AND ATMOSPHERIC HEAT LOSS INTO SPACE INDICATE THE ICE AGE IS ON OUR DOORSTEP.

It took me many months to research and write this book.

During that experience I came to realize how really really difficult the new ice age is going to be.

The **best** we can hope for is a LITTLE ICE AGE in which the society will collapse, and about half the population will die from lack of food and water and the cold. That's the least damaging outcome possible, and even that optimal case that is catastrophic.

Lack of food, lack of heating, lack of electricity, it's going to degenerate countries into a barbarian battleground of neighbors against neighbors.

We can, however, lessen the damage by preparing for the coming ice age. But the demonic Global Warming fraud scientists and politicians, pushing their brainwashed mob, are doing everything possible to stop us from doing so.

The collapse of society and the economy the least awful outcome we can hope for if we do not start **NOW** making massive preparations for the coming freezing weather.

 In the Little Ice age from 1300 to 1800 in Europe and North America, about 10 percent of the population died. Mostly from lack of food.

They were tough, rural and did not depend on electricity and transportation, heating and public water systems to live. We depend on all those things. And these things are going to collapse in urban centers and many other places even during a Small Ice Age. Particularly if we do not prepare infrastructure for the coming cold weather.

We can see clearly what will happen, from what did happen during the previous ice age 1300 – 1850. Growing seasons were shorter, and less sun in those short seasons. People starved. Society collapsed.

Many simply died of hypothermia, they froze.

Today, as mentioned, we are more susceptible to disaster caused by freezing weather than in 1300 to 1800. We've got the modern life based on electricity and highways and roads which we can actually drive on.

During even a small ice age, elevators won't work in our city buildings. Water won't flow through the frozen pipes. Electrical power plants will freeze to a standstill. Trucks to deliver what food there is won't make it in the cold temperatures, and the roads being filled with snow and ice.

So in the best possible case, in a Little Ice age, about half the country will survive. For the survivors, life will be rather unpleasant in a mini-ice age.

If there is a full ice age, and it gets colder and colder and massive ice sheets begin to form, and snow is everywhere, human beings all die.

I suppose the worst news is, according to very strong scientific evidence, the next ice age is about ready to start. We just hope it's a Little Ice Age.

The solar science, which is well developed, has shown that the sun is entering a "Minimum" period. When the sun gets cooler, we get a lot colder.

While completing this book, some major dramatic research discoveries have occurred, which makes it clear that the Ice Age is upon us.

One of them has to do with ocean currents.

Ocean circulation in North Atlantic is at its weakest for 1,500 years - and at levels that previously triggered a mini Ice Age, the study warns

- Currents have a 'profound effect' on both North American and European climate

- Researchers found a similar weak signal during a period called the Little Ice Age, a cold spell observed between about 1600 and 1850 AD

Researchers studied the Atlantic Meridional Overturning Circulation (AMOC), the branch of the North Atlantic circulation that brings warm surface water toward the Arctic and cold deep water toward the equator.

The research, by Drs. Christelle Not and Benoit Thibodeau from The University of Hong Kong, is interpreted to be a direct consequence of global warming and associated melt of the Greenland Ice-Sheet. But what started out as perhaps another Global Warming support paper, rapidly turned against it.

So the first part of the study is of course part of the Global Warming propaganda effort. But what is important, is that there is a link between weak ocean current signals and little ice ages.

Slower circulation in the North Atlantic can yield profound change on both the North American and European climate but also on the African and Asian summer monsoon rainfall.

'The AMOC plays a crucial role in regulating global climate, but scientists are struggling to find reliable indicators of its intensity in the past.

'The discovery of this new record of AMOC will enhance our understanding of its drivers and ultimately help us better comprehend potential near-future change under global warming' said Dr. Thibodeau.

Although this study is just another part of the Global Warming propaganda effort, it is valuable because it points out the weakness of the ocean current-flow. A weakness clearly associated with ice ages.

The research team found a weak signal during a period called the Little Ice Age (a cold spell observed between about 1600 and 1850 AD).

https://www.dailymail.co.uk/sciencetech/article-6430691/Ocean-circulation-North-Atlantic-weakest-1-500-years-trigger-Ice-Age.html

Earlier, in 2009 I wrote a technical paper explaining the technical fraud involved in the Global Warming religion.

And the last part of that paper, I mentioned that because a strong solar minimum, meaning a very weak period of the sun, was due to happen, based on all our solar science, that an Ice Age was on the way. At the time I felt it could start perhaps in 15 years as a rough date-based guess, based on the best calculations I could do relative to the sun entering the next minimum. That would have been 2024. Although it's of course impossible to pinpoint the actual years. Still we can make scientific estimates.

I also mentioned that Russian scientists, who are often very good by-the-way in highly defined fields of advanced work, know Global Warming is bunk and that an Ice Age is on the way.

That's tens of thousands of university professors and other experts. In China it's the same. They do nothing to act against Global Warming because they know it's a sort of evil fairy tale.

When I read the latest ocean current research in late November 2018, and also about the results of the world's most accurate solar model, I had the feeling that I had estimated the date wrong. It would not be 2024, but either sooner that the new Ice Age begins. Possibly this year, 2019 or in 2020. Or according to the world's most accurate solar model, sometime around 2035.

The 2019 date comes from amazing data originating in the NASA Langley Research Laboratory. That is mentioned later in this book.

Thus we are looking about entering the new ice age sometime between 2019 and 2035. This is quite certain and unlike the hysterical "sky is falling" prognostications of Global Warming. Those are based on failed models and fraudulent data.

And I will just add here briefly a bit more about NASA. That in Chapter 22, we cover another very very solid scientific observation which shows the ice age is about to begin. This observation made by the Langely Research Laboratory of NASA. A heavy hitter. So that we save until Chapter 22.

Realizing that the ice age is really here, at our doorstep, the feeling isn't a happy one. I know to some extent what Ice Ages are, both from my cold weather technical experience and studying the history of the most recent ice age between 1300 and 1800.

According to those experiences, the best we can hope for is a disintegration of our society, collapse of our food production, and energy production, and transportation. In the best possible case, only half of Americans will survive possibly much less.

Those who survive will be in a changed and very unpleasant world. Something like the Hollywood films about the future when society had totally disintegrated and violence rules the day. Hollywood is awful on politics, but some of the films about the future could be strangely accurate.

Excuse me for getting personal, but for me, knowing that the Global Warning movement, which has brainwashed the children and students in America, and many other groups here as well.... is stopping our country, stopping our government, stopping our President from taking the necessary huge steps forward to prepare for freezing, is disgusting.

The Global Warming & Climate Change from Human Activity false religion is simply put, the enemy of life, liberty and survival.

The enemy of every single American. And of every single human and animal and plant on the planet.

We must never give in. And at the very least, hope for a Little Ice Age, and act so that the country prepares for it as quickly and as effectively as possible.

President Trump is right, there is no Global Warming and no Climate Change caused by human activity.

The irresponsible Democrat Party, Justice Democrats / Democratic Socialists and Democrat Media Complex, which has hoodwinked much of the country, is driving us to national suicide by ice.

Do not help them. Do not support them. Act against them politically, judicially, and in the media, as if your life, and that of your family and nation, depends on it, because it does.

CHAPTER 25

WHAT ABOUT SOCIAL JUSTICE?

A focused and serious look at the Global Warming & Climate Change fraud, shows that in addition to the money motivation behind it, there is a political aspect. In fact the Climate movement is more political than anything else. Basically it's about power.

Global Warming hysterics' political war has two sections, two platoons:

One relates to the social justice warrior wanting to bankrupt the western high tech industrialized nations, and send the money to the developing world. To the third world.

To use Global Warming hysteria as a hammer against the Western industrialized countries. In short to take them down a peg. To turn advanced societies third world country such as Velezuela.

This (and money) was a central motivation of the UN when they began to push this false religion. Climate and science actually had nothing to do with it. The UN simply wanted to transfer wealth from the West to developing countries.

Put bluntly, the elitist Leftists at the UN planned to damage and destroy the economies of the advanced industrial companies, and send their wealth elsewhere by force.

That's what is behind Carbon taxes and forced purchase of carbon credits by Western industry and society. An effort to cripple and destroy Western high technology societies.

It's sickening and true. And the propagators were mostly from the developed countries they are trying to destroy.

This original UN effort and idea has poked its head up recently in the US.

It's been proclaimed by the new Democrat House of Representatives, and most vocally by Justice Democrats / Democratic Socialists, that somehow by worshiping the false religion of Global Warming, then in the United States there will be justice, racial equality, economic equality, everyone with a job and so on.

Let's take a look at this idea. Which actually is quite insane.

Two of the loudest proponents of this are hard core communist Bernie "the Old Red"

Sanders, and his younger associate, actress playing the role of Socialist, Alexandria Ocasio-Cortez. For ease of identification, we will call Bernie, "The Old Red" and Alexandria, good naturedly known as "AOC".

Both of them are very confused individuals.

When Bernie had to decide where to spend his honeymoon, his choice was of course... Moscow.

But where to start with the other leading member, the face of the Global Warming fraud, who we will refer to with all due respect as AOC.

First and foremost, AOC is an actress. Not a politician. Not even a socialist. As a student at Boston College, according to student publications, she was writing conservative economic papers based on Edmond Burke and others.

That she is an actress, controlled by far left radicals and far left radical organizations, is well documented ON VIDEO. See it for yourselves dear readers on a YouTube investigatory station (called "Mr. Reagan") at:

https://www.youtube.com/watch?v=1h5iv6sECGU

Then her brother submitted her name, as a candidate for the ultra-radical JUSTICE DEMOCRATS. A group managed in part by Zach Exley, Cenk Uygur, Becky Bond, Soros financed organization, and their colleagues in subversion.

The dark money showed up in AOC's congressional campaign in the Bronx. Evidently it had around $780,000 in cash and more than million stashed away (without reporting it to the FEC) in a PAC run by supposedly by AOC and and her campaign manager

According to management of the JUSTICE DEMOCRATS in the amazing investigative video, AOC was chosen from among 10,000 applicants !

YIKES.

If that's true, then there are 9,999 other candidates who were stark raving mad (according to experts, of course).

That having been said, there are a few things AOC and Bernie have said about Global Warming / Climate Change, that would be good to point out.

This is not meant to pile on AOC with criticism. But she has to get her ideas back into the realm of reality and break free of her puppet masters and choose her friends more wisely. Some of her JUSTICE DEMOCRAT pals are racists and antisemites. If she gets back to the conservative roots of her family, and switched party to Republican, she has a bright future.

Here we just mention AOC in relation to the social justice fraud, since she in December 2018 launched the theory in the mass media, that shift to green energy to avoid Global Warming will somehow create a mass of jobs, and end racism, and end evidently everything bad in this country.

But about AOC... she follows the Democrat Party cookie cutter mold and has a tendency to tell "little white lies"... just like every other Democrat, Justice Democrat / Democratic Socialist.

Little white lies, or even whoppers, in order to achieve their objective. Which in the end is always... power. And tyranny.

And once they get that power, the lies multiply along with tyrannical aspects of their behavior. AOC fits the mold perfectly. But remembering that she is just an actress, maybe she can break free.

To those of you on the Left of the political spectrum who are reading this book, and don't know what "little white lies" means because it's a phrase not used very often among you... I will explain that it has nothing to do with racism, or anything like that.

It simply means that the lies are more on the cute side than the ugly side of things. They are more little fibs than lies.

For example, someone who steals a bicycle, or someone's purse or wallet, and who lies and denies it, that is a big ugly lie. Not a little white lie.

But if someone asks a family member to stop off on the way home from work and buy a pound of apples, but the person coming home from work is tired and doesn't do it....

Then a little white lie comes into place: "Oh, I really wanted to get the apples for you, but I completely forgot. The traffic was so busy I was concentrating on that."

That's a little white lie. It's not meant to hurt anyone.

Anyway, among AOC's campaign little white lies were:

- That she was a poor girl growing up in the Bronx

    Actually her father was a financially successful and brilliant architect, and she grew up in the wealthy suburb of Westchester County.

- That she spent her youth working as a waitress, and never mentioned any formal education after high school.

    In reality she attended and graduated one of the best higher education centers on America's East Coast, Boston College.

    One of the leading colleges in the US, and it costs about $52,000 a year to go there. So on four years after getting her degree the family plunked down $200,000. Maybe it was a little less when she went, but still it was still among the most expensive colleges or universities in the US.

- Again the waitress as a career story

    In fact she was a staffer to Senator Ted Kennedy

- That in her primary campaign for the Democrat congressional district in 2018, she won by knocking on doors herself and meeting with the people. Bless her little heart.

    In fact she had over 200 full time volunteers, Democratic Socialists, who mostly came from outside the district and did the knocking for her for many months. They received housing and food and probably spending money as well.

    Their work cost perhaps a few hundred thousand dollars.

- The next lie... and, the pile is getting higher and higher...

    That she won the primary campaign with no real campaign budget or funds, just knocking on doors all day, every day

    Knock, knock, poor me. Except that she actually had about $780,000 in campaign funds for the primary campaign. Perhaps in large part from millionaire leftist mega-donors.

    And apparently a million more stashed away secretly in some PAC which she and her campaign manager operated.

So, there was a pile of little white lies.

And not really her own. Remember, she's just an actress. They were written by the brains behind the JUSTICE DEMOCRATS.

Lies are what Democrats and particularly the Democrat Party, Justice Democrats / Democratic Socialists do always in order to get elected.

Well, little white lies in their own eyes, but in reality they are really whoppers.

WHOPPERS FROM THE SECRETIVE BEHIND THE SCENES JUSTICE DEMOCRAT ORGANIZATION

The purpose of this organization is none other than to sabotage and destroy the democratic process of elections in the USA. So as to flood the field with well financed JUSTICE DEMOCRAT candidates.

To replace first the Democrat Party candidates. And then later to do the same by inserting fake candidates into the Republican primary elections. Total destruction of the free and fair election system is their objective. Just as it is the objective of every totalitarian movement.

AOC is just a pawn in this diabolical program of subversion against American liberty and inalienable rights for every American.

Puppet masters are pulling the strings. Reportedly in terms of management this includes but is not limited to Zach Exley, Becky Bond, and at least at the start, Cenk Uygur.

But they themselves are apparently puppets of dark shadow Left wing billionaires who finance the destruction of American democracy.

JUSTICE DEMOCRATS

First, it's useful to note that "JUSTICE DEMOCRATS" is simply a tricky name. The BIG LIES from the political Left are always based on words which mean the opposite of what they appear to mean.

They are one in the same as the "Democratic Socialist" Party, but have decided not to use that name now, because it might scare voters.

But even the "Justice Democrats" name is a fraud. Nothing could be farther from the so-called JUSTICE DEMOCRATS, than justice.

They stand for racism, antisemitism, tyranny and destruction of the rights of American citizens as clearly stated in the Constitution. As we will see, "justice" figures into their program only as a target. They seek to destroy it.

As I mentioned, another name is "DEMOCRAT SOCIALISTS". AOC was a candidate under that name although she is now featured on the JUSTICE DEMOCRAT website as their candidate.

In local city elections in March and April, such as in Chicago, many candidates for city councils ran under the "SOCIALIST DEMOCRAT" name and won.

It's a clever trick. The names makes voters think they are in the Democrat Party. They are not. Their program actions clearly demonstrate that they intend to destroy it.

These two new political parties (really one party, two names.. Justice Dems and Dem Socialists) are running on the same BIG LIE. The 3 political talking points developed by Exley and Bond. Like the Nazi propagandists of Hitler, who they apparently copy and emulate, they keep it simple, and of course fake:

- **Medicare For All...**
  Meaning free health care for all. Simply impossible of course. And something which will destroy the entire US healthcare system, and make the country like Venezuela, Cuba, or North Korea. Healthcare For None. Except for the tyrannical JUSTICE SOCIALIST / DEMOCRATIC SOCIALISTS leaders of course.

- **The Green New Deal...**
  Which actually means destruction of the US economy, massive poverty, homelessness, disease and unemployment. And it won't effect climate one bit. It's 100% a power grab.

- **Social Justice...**
  Their code words for their own racism. If you are an ordinary, what they call "white" American, you are are in for a life of discrimination and being targeted.

  Instead of justice and liberty for all, it's power to destroy for a few select minority groups to destroy the "whites". Why not make this clear. This is a racist and bigoted party.

  The bigots are against Christianity, against Judiasm. Let's repeat this: The Justice Dems / Dem Socialists are racists, anti-Christian and antisemitic.

  That's their "social justice".

These are like the three big lies of the Russian Revolution: Land, Peace, Bread

It was simple. Sounded good.

Or the same as the Nazi big 3: One leader, one race, one empire.

The results of the big three however were in Russia: NO LAND FOR THE PEOPLE, NO PEACE FOR THE PEOPLE, NO BREAD FOR THE PEOPLE.

In Germany the results of the big three lies: insane leaders who in the end committed suicide and took the country with them, blue eyed blond hair aryans nowhere to be found, a bombed out country with people homeless and starving.

Now we look at the present day BIG THREE LIES, these of the Justice Dems / Dem Socialists:

NO MEDICAL CARE AT ALL FOR THE PEOPLE

NO CLIMATE IMPROVEMENT BUT CERTAINLY LOSS OF LIBERTY AND RIGHTS

AND RACISM AND OPPRESSION BASED ON SKIN COLOR AND RELIGION, SPREAD THROUGHOUT THE LAND. PERHAPS LEADING TO CIVIL WAR.

None of them, not one, not even little ol' AOC, give a hoot about climate.

She of course excused herself for using SUVs as transportation when the subway was right there, for flying all over the place, for using air conditioning and so on. Would someone worried about climate destruction do that?

JUSTICE DEMOCRATS / DEMOCRATIC SOCIALISTS (and in particular the puppet-masters behind them) want one thing... power. The Democrat Party is the same.

And the Justice Democrats / Democratic Socialists seek to gain that power by Pearl Harbor attacks.

Inserting candidates who support racism, anti-semitism, and destruction of the democratic political and election process into the elections. While hiding their true intent and beliefs, they masquerade as "moderates". As fake Democrat Party members.

Regarding climate, the "Green New Deal" is their baby. The only thing green about it, is the grab for money. That money used to gain totalitarian power. The so-called "Green New Deal" plainly and simply a way to grab power.

A better name would be the "Green New Steal". Theft of power. Theft of money.

The bible says that Satan comes as an angel of light. And all the JUSTICE DEMOCRATS are sweetly smiling. Until they get elected. Then all hell breaks loose.

Justice Democrats in the House

2019 Congress kicks off with 7 Justice Democrats pushing for change in the U.S. House of Representatives.

- Alexandria Ocasio-Cortez — NY-14
- Ayanna Pressley — MA-07
- Ilhan Omar — MN-05
- Pramila Jayapal — WA-07
- Rashida Tlaib — MI-13
- Raúl Grijalva — AZ-03
- Ro Khanna — CA-17

THE SO-CALLED JUSTICE DEMOCRATS ARE IN FACT NAZI-TYPE ORGANIZATIONS. BASED ON NAZI PROPAGANDA METHODS. WHICH THEY HAVE APPARENTLY STUDIED IN GREAT DETAIL.

Their smiling photos above mislead about why and even how, they have been elected to congress.

The Nazi propaganda minister, Joseph Goebbles explains way:

"We enter congress in order to supply ourselves, in the arsenal of democracy, with its own weapons.

If democracy is so stupid as to give us free tickets and salaries for this bear's work, that is its affair.

We do not come as friends, nor even as neutrals. We come as enemies. As the wolf bursts into the flock, so we come. "

They are the big lie from beginning to end. Masterminded by former Bernie Sanders media team Zach Exley and Becky Bond, their programs are Halloween horror stories dressed up as a princess. It's all lies. Impossible lies with intent to deceive:

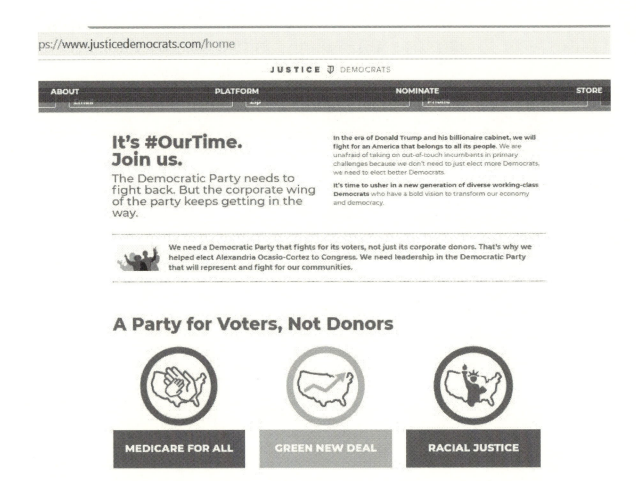

There is no difference between these lies and those of Stalin or the Nazi used to destroy democracy in their countries.

EXLEY AND BECKY BOND ARE NOT THE MESSAGING GENIUSES THEY PRESENT THEMSELVES TO BE.

THEY ARE SIMPLY FOLLOWING THE NAZI METHODS.

FOLLOWING THE TEACHINGS OF THEIR MASTER, JOSEPH GOEBBLES, PROPAGANDA MINISTER OF THE NAZI THIRD REICH:

"If you tell a lie big enough and keep repeating it, people will eventually come to believe it.

The lie can be maintained only for such time as the State can shield the people from the political, economic and / or military consequences of the lie.

It thus becomes vitally important for the State to use all of its powers to

repress dissent, for the truth is the mortal enemy of the lie, and thus by extension, the truth is the greatest enemy of the State."

All three of the Social Justice policies are the BIG LIE. And the Green New Deal, which is based on the Global Warming & Climate Change from human activity lies is a major pillar of their activities. Their purpose is basically to destroy the country. And then install their totalitarian state.

Social Justice, for example, is based on the insane policy of OPEN BORDERS. Something which of course will lead to collapse and chaos with the country flooded by millions and millions of new illegal immigrants every year.

But because the Justice Dems / Dem Socialists hate America, this is their objective. Catastrophic disaster which their own policies cause.

MEDICARE FOR ALL, in reality means Medicare for none.

Medicare is almost bankrupt now, and basically through their program, the American health care program will collapse, and not be available to anyone.

Let's repeat again the words of the people who inspired Exley and Becky Bond, the brains behind this JUSTICE DEMOCRAT movement:

We enter congress in order to supply ourselves, in the arsenal of democracy, with its own weapons.

If democracy is so stupid as to give us free tickets and salaries for this bear's work, that is its affair.

We do not come as friends, nor even as neutrals. We come as enemies. As the wolf bursts into the flock, so we come.

This attempt at the secretive overthrow of American liberty, freedom and democracy, fits the Global Warming / Climate Change From Human activity movement, because it is one HUGE lie. A lie with a political purpose: to grab absolute power.

IS IT A BIT TOO MUCH TO POINT OUT THE SIMILARITIES BETWEEN JUSTICE

DEMOCRATS AND THE NAZI PARTY ?

OK, let's look at a few of them. They all basically fit the same cookie cutter mold, but we just look at a few now.

AOC lied her head off to get elected. The supposedly "poor girl from the Bronx" was actually a well-to-do person from the wealthy suburb of Westchester. Instead of "knocking on doors" as she claimed she did, 200 to 250 trained Dem Socialists volunteers, most from out of state, worked for months knocking on doors. They were provided room and board and spending money. And how about the claim she financed the campaign, and it was grass roots? Noooo... $780,000 in campaign funds plus evidently another million dollars hidden in a PAC. Dark money from the Leftist billionaires.

And then to follow the lies typical of all Democrat Party / Socialist members used to get elected, she after being elected to, and joining the House of Representatives, unleashed the tyrannical side of the movement.

Remember she's an actress, not a politician. But maybe the power is going to her head? Maybe we started with a play acting Frankenstein political monster. But before our eyes, she and all the other JUSTICE DEMOCRATS are morphing into ever growing monsters of tyranny. Shutting up anyone who gets in their way.

It's not going to far to say, that the Green New Deal is a Nazi ideology. Remember it's not really about climate. It's about power.

Brought to us by tyrannical misfits.

Here are the gruesome duo together. Zach Exley and Becky Bond. Modern day Nazi style propagandists. Joseph Goebbels would be proud of his two students. They are always smiling in photos. Smiling until they get some actual power. Then the evil starts pouring out of those mouths:

On December 7, 2018 (coincidentally Pearl Harbor attack day), AOC threatened Donald Trump Jr. with a congressional subpoena because he trolled her in a tweet.

The tyrannical Soviet side of radical Democrats (and aren't they all radical now) always comes out. In this case, even before she sat down in her seat in congress. Yikes.

Shortly after that she threatened members of the Democrat Party, she was "keeping a list" of all who did not support the radical JUSTICE DEM agenda. And those people would be primaried by the totalitarian Justice Dem candidates in the coming elections, thus kicked out of congress.

Inside every Democrat Party member or Socialist, or Justice Socialist, there seems to be a little Stalin, a little Soviet totalitarian, just screaming to get out. A little Joseph Goebbels.

THE JUSTICE DEMOCRATS, vipers spawned by the Democrat Party use Nazi methods. This is how they work, and the GLOBAL WARMING lie, and Green New Deal is one of their main tools.

This book is about climate. But we are talking a lot about politics. Why?

Simply because the Global Warming movement, the Climate Change From Human Activity movement, are not climate movements. They are political movements.

And the agenda of those puppet masters behind these movements, is to seize absolute power.

Make no mistake about it, some of the worst Nazi tactics, such as rounding up political opponents and imprisoning them, torturing them, killing them, are not happening yet by Justice Democrats for basically two reasons:

The US Constitution.

The Second Amendment.

It's for certain that the leadership and puppet masters fear these two things. The protection of liberty by the Constitution. And they fear coming face-to-face with Mr. and Ms. Second Amendment.

So instead of the prison, torture, and death, they stick to the Nazi type propaganda, with the goal of suppressing freedom:

If you tell a lie big enough and keep repeating it, people will eventually come to believe it.

The lie can be maintained only for such time as the State can shield the people from the political, economic and/or military consequences of the lie. It thus becomes vitally important for the State to use all of its powers to repress dissent, for the truth is the mortal enemy of the lie, and thus by extension, the truth is the greatest enemy of the State

As soon as possible, however, the Justice Democrats and in fact the Democrat Party, will move from words to tyranny.. at the first available opportunity.

The Democrat Party, and their viper children, the brains and billionaire donors behind the JUSTICE DEMOCRATS / DEMOCRATIC SOCIALIST have more in mind than just their climate lies.

They have in mind what Goebbels referred to as "repression of dissent".

This is why the Democrat Party, the JUSTICE DEMOCRATS / DEMOCRAT SOCIALISTS, as we mentioned above, hate the Constitution of the United States.

AOC, Ilhan Omar, and Rashida Taliab are examples of people who will not tolerate dissent against their ideas. Ideas which often are based on racism, totalitarian ideology, and antisemitism.

The Constitution, which the billionaire donors and the brains behind the JUSTICE DEMOCRATS / DEMOCRATIC SOCIALISTS HATE, is the supreme law of the land, promises and protects the right of dissent. That is why they hate it. It gets in their way.

In so far as the Constitution protects what they have to say, they will use it. But as far an any opposition political party or We the People, they are working day and night to remove the right of dissent.

As we read in the words of Joseph Goebbels, their lies will collapse sooner or later.

And at that time, the only way to keep control is by more repression of descent.

This is the reason the Left. The Democrat Party, and the Justice Democrats are so strong on taking guns.

Here is good ol Becky Bond in front of her "STOP the NRA" poster, which really means... disarm the citizens so they become helpless. Stop the people from defending themselves against tyranny.

On that wall are three crazy ideas:

Nuclear power supplies about 20% of American electrical energy. So just shut it down, right? And homes, business, industry, schools, hospitals, have no electricity:

Second JUSTICE DEMOCRAT NAZI idea: disarm the people.

Third propaganda poster: Support abortion. Yep, 18 million African American babies killed by abortion. Is it really too much to compare these people to the Nazi movement?

About the NRA, Becky and Zach, deep in their little hearts, could care less about shootings. They simply are afraid that if they try to crush We the People, now, they will meet armed resistance.

Totalitarians always try to disarm the people. After they disarm the people, the totalitarians shoot citizens.

All those who are in favor of gun control, please raise your right hand

But if the Democrat Party, and their little vipers the Justice Democrats, manage to shred the constitution, and take guns, then the GLOBAL WARMING lie will become a way of life for the people. No transportation, no heat in winter, no food. Except for them.

The totalitarians always keep their guns. Keep their cars. Keep their jets. They intend as the elite to rule over the masses.

Hitler warned against allowing any possible opponent factions or groups to have guns.

The inner totalitarians inside Democrats and so-called Justice Democrats, do get out. And start doing all kinds of nasty things.

And the Democrat Party and the Justice Democrat / Democratic Socialist vipers nest it has spawned becomes more clearly every day totalitarian.

Now of course... Zach and Becky can say whatever they want.. They have those rights under the Constitution they would like to shred.

Just saying, that like some low budget Hollywood horror film, their seven little puppets will keep morphing into little tyrants, and one fine day, bite their heads off. Just saying.

And speaking of biting. And adding, biting the hand that feeds you, how could we leave this section of this wonderful chapter, without mentioning our friend (and they're all are friends, right?), little ol' Ihlan Omar.

Biting the hand that feeds you, means, for those college and university students who have been brainwashed on their campuses by the political Left, it means that someone helps you and you don't show any thanks or gratitude. Not only that you attack them. Biting the hand that feeds you.

Ilhan Omar and her entire family were brought to the US from war-torn Somalia. At the expense of the American people. Given safety, a home to live in, food, education, most expenses. And of course as a radical islamist (and this is said with all due respect, but it's accurate and she will be the first to say so), she does not like America. Not only Trump's America. But any America that is not islamic.

The point here is about the totalitarian way of life that pervades the Democrat Party, and the so-called Justice Democrats / Democratic Socialists.

Ilhan Omar would not be in congress were it not for the large Jewish community in her district in Minnesota. They are almost entirely Left wing Democrats. They trained her in the Democrat Party authoritarian way. Without their votes, she would not be in congress.

The Somali and Jewish communities in that area of Minnesota get along well.

As a bit of history, the Somali people coming, courtesy of Barak Obama, who brought 25,000 into the country. And they brought another 50,000 by the chain migration regulation craziness, that a family member can bring in their families from abroad as well.

It's nice they found safety. Most Somali people and Jewish people in Minnesota get along well, and have built bridges between their communities. Until Ilhan Omar unveiled her totalitarian agenda. The Jewish community that helped her, that put her in congress, is being bitten now. Ouch. They are wondering, "What was I thinking?"

Totalitarian Left wing Democrat Party members, and totalitarian radical islamists get along very well. Up to a point.

That first point is, the Jews in the Democrat Party. Intolerable of course.

And the second point is, that in the end, ALL THE DEMOCRATIC SOCIALISTS who are not islamic, become second class citizens in an islamic state.

So Zach and Becky, watch your backs. It's a totalitarian movement you are in and you are helping to create.

Totalitarians always eat their own children.

THE GREEN NEW DEAL MEANS NOTHING RELATED TO CLIMATE, BUT IT DOES MEAN CONTROL OF FREE SPEECH

The Global Warming / Climate Change Caused by Human Activity movement is of course totalitarian.

Trying to control thought, speech, and every aspect of lives in America. Liberty and justice will be taken away from the American people on the grounds of a Global Warming emergency. IF THE DEMOCRAT PARTY and its little vipers of "freshmen" congress people get their way.

The idea that AOC would launch a personal vendetta, against not only Donald Jr, but against anyone, about something someone exercising their free speech said about her in a tweet, or keep a list of her Democrat colleagues who oppose her and other Justice Democrats, tells you everything you need to know.

The puppet masters of AOC are totalitarian monsters, financed perhaps by the equally crazed Left wing billionaires.

A PHOTO I DIDN'T WANT TO PUT IN THIS BOOK.

BUT HERE IT IS...

Following is a photo I did not want to include in this book. But sadly it's necessary.

It has to be included, because about 10% of the readers of this book will be members of the political Left.

Some of them may read the ideas and statements of Joseph Goebbels and get all excited about them. Excited in the sense of thinking, how cool is that!

Thinking these propaganda methods are something great that the Democrat Party, Justice Democrats / Democratic Socialists can use even more than they are doing now.

Therefore here is the photo. It shows the end of the road anyone using Goebbels ideas will end up on.

On May 1, 1945 Joseph Goebbels murdered his six children.

Five daughters and a son. All small children, the oldest being 12 years of age Helga Susanne.

Goebbels murdered them by giving them cyanide capsules.

This is a photo of a Russian military doctor in a post autopsy on Helga Susanne, age 12. Shortly after her death.

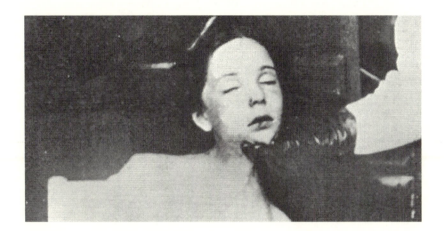

Helga's Susanne's face had multiple bruises. Leading the medical specialists to believe she had struggled against being given the poison.

This result, Democrat Party, Justice Democrats / Democratic Socialists is the path you will be going down when you utilize the methods of Joseph Goebbels for propaganda, and political power.

If the first photo didn't grab the Goebbels fans, maybe this one, from a little different angle will:

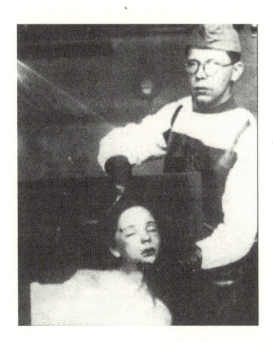

And hey, it would be good to mention that Goebbels did not kill just his six children.

His BIG LIE resulted in the deaths of millions of Jewish children, Polish, Children, Russian Children, German Children, disabled children, Hungarian children, French Children and on and on.

That's the road's of totalitarian propaganda and lies. The road of Goebbels. The road which the Democrat Party, Justice Democrats / Democratic Socialists has been following the script of, fairly exactly up until now.

Like most Democrat Party / Justice Democrat operatives, whether in congress, or in the Democrat Media Complex, or on the Special Counsel of Mueller, these political radicals are dangerous to freedom if we allow them to slash and burn through the pathways to power without political challenge.

Without political defeat. The elections of 2020 and 2022 are right around-the-corner.

I wonder if AOC will threaten to subpoena me for writing these truthful things about her. I'm a member of the Press. An investigative reporter. The video about her exists:

https://www.youtube.com/watch?v=1h5iv6sECGU

Zachary Exley and Becky Bond, the propaganda ministers of the JUSTICE SOCIALIST group, channel the Nazi totalitarian and propaganda ideology quite amazingly. How can they deny it?

In writing this book, I was surprised how the Nazi propaganda machine and the JUSTICE SOCIALIST / DEMOCRAT SOCIALIST propaganda machines work in similar ways. The JS / DS are almost a carbon copy of the Nazi propaganda methods.

All of them have essentially trapped themselves:

"The essence of propaganda consists in winning people over to an idea so sincerely, so vitally, that in the end they succumb to it utterly and can never escape from it."

Goebbels...

AGAIN WE ASK, ISN'T THIS A BOOK ABOUT CLIMATE?

Well, it is.

But for the Democrat Party, Justice Democrats / Democratic Socialists, the Global Warming and Climate Change movement is not about climate. It's all about a way to control people. It's all about power.

None of them cuts their jet travel, their SUV jaunts when subways are a minute walk away, or turn off their air conditioning.

They don't give a crap about the climate.

It's all part of the BIG LIE to size power.

Again about AOC.. Who as we know is an actress... now apparently morphing into a totalitarian.

Does she believe in freedom of the press for CNN, for MSNBC? Good, then she certainly believes in it for me a certified member of the press. She started as an actress, and hopefully hasn't morphed irrevocably into a concrete version of a totalitarian Frankenstein monster.

And I believe that finally she will do the right thing and not even send me a nasty tweet. Stand up to those puppet masters AOC !

These comments are not means to pile on her career. They are in fact an invitation to turn right into the light. An official invitation to become a Republican.

Leave the darkness and come into the light AOC.

While a student at Boston College she wrote some papers in economics which were conservative and also brilliant. These were published in Boston College student publications. Reading them we see a very intelligent person and a conservative.

What happened between then and now is unknown, but the papers show there is hope.

Please read your own papers again. Please listen to a recording of "Amazing Grace", sung by Judy Collins, every day for a few weeks.

Break free of your puppet masters.

I believe then you will make the shift to become a Republican. The Republican Party is the party of Freedom. The real party of Justice.

And the Republican Party will welcome the new conservative you with open arms. As we do all who want to join with us in the fight for all American citizens' rights and freedoms.

There are many benefits to becoming a Republican. Because then she can be elected by telling the truth. Wouldn't it be nice to stop lying?

She can set up (and I am willing to help her after she becomes a conservative constitutional Republican) solid, real alternative energy programs that fit economic realities and actually succeed.

I'm an engineer working in that field of alternative energy for over 25 years, with many international projects. From hydropower, to wind, to solar.

And she can stop lying. Lying is not a good life.

Democrat Party, Justice Democrats / Democratic Socialist operatives always lie in campaigns. And after they are elected to congress.

If they told the truth they would never be elected. And none of their laws would ever be enacted if the truth were known.

GLOBAL WARMING and Climate Change Caused by Human Behavior itself is a lie and it has brainwashed part of a whole generation.

Global Warming / Climate Change is used as a tool because the Democrat Party wants power and will do anything to get it.

So hopefully she, and infact ALL OF THE SO-CALLED JUSTICE DEMOCRATS, will get out of the Democrat rut, and show the world how really wonderful a person they are.

Come on all of you. Come into the light. Join the party of Freedom. You too, Zach and Becky.

AOC's mother got out of Democrat New York City and she can too.

Welcome on board Alexandria, Zach, Becky and all of you... after you realize that Global Warming is a wicked fairy tale, and that an Ice Age is coming very soon.

And we need to get our infrastructure ready for it.

BUT WHAT ABOUT SOCIAL JUSTICE THROUGH THE GREEN NEW DEAL ?

ONE OF THE JUSTICE DEMOCRAT / DEMOCRAT SOCIALIST BIG LIES

Now lets focus under the microscope, if the Democrat cookie cutter energy and funding policies of the Green New Deal would actually ever work.

Let's start with the crazy idea that somehow shifting to green energy by force of the government would provide massive jobs, end racism, and bring social justice.

Actually Barak Obama tried that. Not social justice, but shifting to alternative energy through taxpayers' money. Or we should say, shifting taxpayers money to Global Warming fraudsters.

Yes they (the fraudsters) grabbed €500 million in just one scam alone and ran.

Solar energy just happens to be part of the work that I do. So I know a little about it.

Evidently because Obama was supposedly going to stop the sea from rising, his abilities were apparently super intelligent. This was proven false time after time after time. He was all in for wind turbines and solar electric systems but really didn't know a thing about them. Except that they were all shiny and new.

So enthusiastic was he about solar electric systems, he authorized a $500 million unsecured loan (yes five hundred million dollars) of taxpayers money to a crackpot new company that had a new deign of solar photo electric panels.

It's name was Solyndra.

It's a pity that AOC wasn't around in politics then, because, she could have used that money to find a solution to cow farts.

Because, you see, dear readers, there were a few little problems with Solyndra. And as we mentioned previously, it shows what disasters would lie ahead in any Green New Deal, in which the energy segment of the economy is run by government.

And now some details about Solyndra. As we all know, solar panels are flat and in the shape of a window or door. That's because the flat shape works.

To some extent anyway. Solar panels are terribly inefficient only about 22% efficient. Really awful.

And after 25 years, they have to be thrown into the garbage. Because they stop working.

Non recyclable garbage. But we will get to that later in this reading adventure together.

And while I mention the adventure of this book, I'd like to thank you, my dear readers.

Without you, there would be nothing. Just ink on paper. But you bring life and hope and action to it.

Well, back at good old Solyndra, it wasn't a flat panel like all the others in the world, it was like a big glass tube. Like those test tubes in high school chemistry, but quite a bit larger.

The round glass pipe solar panel was then just a few inches high and very very long. Many many yards long. A long glass pipe. So far, so good. Well it seemed that way at first...

The Solyndra solar photo voltaic equipment was sort of like a transparent glass pipe about 12 inches in diameter. I've never seen one up close, just photos, and as far as I know many are in the garbage now, because they overheated. So I'm guessing the diameter of the glass tube from photos.

To keep birds, animals, dirt, dust and insects out of these genius new type solar panels, the glass pipe had to be sealed. With the exception that the electric wires could get in and out of the tube through air-tight holes.

But there was also a teeny weeny little problem with this technology. Which we see here in the long glass tubes sitting somewhere.

Any first year college student of alternative energy knows that the hotter it gets, the less efficient solar panels are. That's why there are no solar electric panels in really hot places like Dubai, or Saudi Arabia, even though they have a lot of sun.

In addition to becoming very inefficient, the panels which become very very hot, sort of self-destruct a little every day, and their product life use is shortened considerably. At most it's 25 years in the best cases of the flat ordinary solar panels. In this case because of the massive build-up of heat inside the tubes, working time might have been reduced to just a few years. I don't know exactly, because I haven't bothered to calculate the temperature inside the glass pipes, but something like that.

Evidently the geniuses at Solyndra didn't know what any child knows, or even house pets know. That on a sunny day, if we stand inside our home behind a window facing the sun, we feel very hot. The heat comes through the glass. Well guess what? The technology of Solyndra, the $500 million company, focused heat into their tubes, just like that.

It became so hot inside the glass tubes that it of course started cooking the solar panels inside. And their efficiency and product life went down the tubes, so to speak.

And there was another teeny weeny little problem with Solyndra, or should we day, the Solyndra scam.

The management and owners of the new $500 million company, were apparently not very good at management or book keeping. So when the US government tried to get back the taxpayers' money, it was all gone. Poof !

As always, the government, filled with bureaucrats involved in projects they have little expertise about, can't do industrial or technical projects very well. It's like having an UBER driver put in the pilot's seat of a military jet fighter. They're not up to the job.

Of course there are many nice and effective bureaucrats who help us every day. But a lot are not, and even more simply don't have the technical expertise to do this type project.

This is a traditional problem of the Democrat Party. They always screw up programs which they are in charge of.

For example, it would be particularly dangerous to put Democrat Party members in charge of your kitchen blender, much less a major technology project. They always get it wrong.

Dear reader, as I mentioned, your author, me, writing this book, happens to be an engineer who has worked in renewable energy for over 25 years. I don't know everything, but something anyway.

Regarding solar energy, the really big projects in solar don't seem to go very well.

That's why I stuck with smaller ones except for hydropower. Hydropower is proven, and can be huge in scale and I did that as well.

But really big solar, particularly those huge solar products that run on the heat of the sun, have problems.

Take for example the Ivanpah project financed by Google. It heats a liquid that by heat transfer creates steam to power turbine electric generators. These gigantic solar energy plants work (or are

supposed to anyway) by focusing the sun by thousands of mirrors in a desert.

A great idea, except that the heat from the mirrors kills birds.

And oh yes, I should probably mention that this $ 2 billion plant doesn't work. It's too complicated. The thousands of mirrors are too heavy, there are computer and motor problems lining up the mirrors on the tower target, and almost endless other glitches. The whole structure and design is crazy.

Oh yeah, and it caught fire too.

I knew well the company that designed and built it. In their office is a big photo of the Ivanpah project. I said to myself, this isn't going to work.

It cost over $ 2 billion. Yes, billion, not million. A lot of that taxpayer money I assume, and it has to run on natural gas much of the time.

A lot of the funding came from the Google environmental radicals and a few other large companies. But they came crying to the government for help when this huge thing wasn't able to produce much electricity from sunlight.

They were asking the US government for taxpayer money so they (Google) could get their money back.

ALTERNATIVE ENERTY CAN'T DO IT.

IT CAN HELP ONLY A LITTLE BIT. ABOUT 12% TO 18% OF WHAT WE NEED.

Alternative energy can't come anywhere near the amount of power needed for American personal and industrial use.

In rare geographic locations, such as Alaska under the amazing work of Governor Sarah Palin, the state reached meeting 50% of its energy needs from alternative energy.

That was in part because of the excellent natural lakes and rivers infrastructure, leading to hydropower system producing massive amounts of electricity. And a low population level and industrial level. So they don't use that much electricity in the first place relative to the rest of the country..

Most states could, even with a huge effort and expense, meet at most 12% to 18% of their power needs by alternative energy (solar, wind, and best of all hydropower).

And it would be hugely expensive. Almost double the cost of fossil fuel energy. Which by the way, thanks to American technology is quite clean now.

The only way alternative energies such as solar and wind could support US power needs is if the Democrats destroy the entire industrial base of the country, and push our people into abject poverty.

So that we become like that of Cuba, Venezuela, North Korea. They don't use much power for industry and personal consumption.

Then after we have no industry at all, and all live in abject poverty, then alternative energy could meet much of the energy needs of America. No refrigerators, no cars, no electrical appliances and equipment, no home washing or cooking equipment. No hospitals. The Democrat, Justice Democrat paradise.

And there is yet another sad surprise, for the believers of alternative energy as a solution to meeting energy needs without fossil fuels.

That surprise is.... ready Global Warming believers... it creates about the same or more CO_2 to make the alternative energy products than they save.

Yes indeed. It takes a lot of energy from fossil fuels to produce solar panels and to produce wind turbines. This equipment doesn't just pop up by magic.

Making a wind turbine creates about the same or more CO_2 than the wind turbine can make up for.

So where does the Social justice come into the Green New Deal ???

It's a Democrat and Justice Democrat crazy idea, that setting up wind turbines and solar electric power systems would somehow effect racism. How it would do is beyond me.

In fact it would make racism worse. Much worse. (I don't know that's true, but it's no more crazy than the idea that it would reduce racism).

And now for the big big problem with this nutty idea that moving to "green energy" would somehow solve social justice problems.

We are heading into a cold period. An ice age. There isn't any social justice in an Ice Age.

The earth always goes through different periods. Sometimes warming, sometimes cooling. Sometimes for a few hundred years, sometimes for much longer. And right now, based on the controlling factor, the sun, we are heading into a new ice age. Very very cold.

The ocean currents are slower than in any time during the last 1,500 years, and that is also a sure scientifically solid sign of the coming ice age.

Heat is radiating out of the atmosphere into space at an alarming rate, according to the NASA Langley Research Laboratory.

And solar scientists and the leading solar model, predict a solar minimum coming between now and 2035. In fact that we are in the early stages of it already.

When the very cold weather hits, there will be a much shorter and a much colder summer. That means less food.

Collapse of the society, its agriculture, its industry, its transportation, is not good for social justice, or racism? Is it?

When the last part of the little ice age hit Europe in the 1700's to 1800s, 10% of the population died for lack of food.

This time, the country is primarily urban. Elevators. Water pumped into apartments. Sewerage systems. Toilets. Cars. Trucks. None of those would work in an ice age. At least half of New York city will die, and probably much more than that.

And here is big news for the Global Warming / Climate Change from human activity nitwits:

Alternative energy equipment won't work in an ice age.

Turbines won't turn and solar panels will be covered with snow and ice. Sunlight will be at low levels.

Alternative energy won't work in hot global warming either, but since we are heading into an ice age, we focus on that.

How are the plans of the hysterical political Left and the Leftist billionaires going to bring social justice?

By their absolute failure to prepare for the coming ice age, they brings the opposite: the disintegration of society into violence.

The only thing that can help significantly in an ice age are the fossil fuels which the Global Warming religionists hate so much.

Global Warming believers are automatically ICE AGE DENIERS.

ICE AGE DENIERS

**YES.. THE JUSTICE DEMOCRATS / DEMOCRATIC SOCIALISTS... THE DEMOCRAT PARTY...
ARE ALL ICE AGE DENIERS.**

Meaning they not only tell the BIG LIE, but they also work against the clear and apparent and scientifically founded truth that we are heading into much colder period on earth.

That they are ICE AGE DENIERS, is a sort of warfare against the United States and its people.

And they are keeping us from getting ready for the cold weather. We are not doing the necessary things to save human lives (or plant and animal life) and to save our country.

Because the hysterical Left has cast a evil magic spell of lies, and forces focus on heat which isn't coming, the Ice Age will destroy social justice forever.

The wicked lies of the Left about Global Warming are like this: a tiger is slowly coming from behind us to attack, but the Left is yelling, look in front of you, a turtle is coming.

Keep your eye on the turtle, they say. Then the ice tiger sneaks up on us and... you can guess. We're tiger tidbits.

And to end this cheery chapter, your author, who has lived and researched in the coldest parts of the world, can tell you with certainty: wind turbines will not turn in an ice age. They will freeze. I've seen it in Havöysund, Norway.

They turbines look beautiful. Gigantic and majestic. But in the winter, they just stand there. Frozen still. They do not turn. Just like half of America will be, because Global Warming lunatics are pushing the country in the wrong direction and we are not getting ready for the cold.

Solar electric panels will be covered with snow and ice. Power lines carrying electric powers will break because of the snow and ice. Electric generators will freeze shut. Roads will be impassible with snow and ice.

So if you really want social justice, which above all means Life, Liberty, and the Pursuit of Happiness, better make the shift. Walk Away from the Democrat plantation slavery of the mind
.
By supporting Global Warming, the tyrannical Left are pushing everyone in the country, including themselves, into national suicide in the coming Ice Age.

The political left are the killers of social justice.

SO....

JUSTICE DEMOCRATS and DEMOCRATIC SOCIALISTS, now isn't it time for you to step up to the plate about the Global Warming / Climate Change From Human Behavior disaster that you proclaim is coming in 12 years?

It is an inconvenient truth, dear readers, that every single one of the climate hysterics, and with no exceptions, every single one of them farts methane and exhales CO_2.

Every single one of the JUSTICE DEMOCRATS / DEMOCRATIC SOCIALISTS is a little chemical factory, which according to their lies, is destroying the planet. All that methane and CO_2 coming from them!

Time therefore now for the Democrats, Justice Democrats and whatever name they give themselves, to follow the advice of AOC: "Put your bodies on the line."

Take a hit for the team.

Take a hit for the planet.

Take a hit for the trees, polar bears, sea otters, and for your fellow human beings.

Stop breathing CO2 and stop farting methane.

Come on JUSTICE DEMOCRATS, come on DEMOCRAT SOCIALISTS, you can do it. Show us you believe in what you say. Show us you are not gutless cowards.
How can you look at yourselves in the mirror every morning, knowing that you are farting methane and exhaling CO2?

Come on. We who look for a bright future for our children, are waiting to see if you will take a hit for the team.

Fortunately, because CO2 and methane have no effect on temperature and climate what-so every. No one has to do that.

But since you JUSTICE DEMOCRATS / DEMOCRATIC SOCIALISTS believe what you say, take a hit for the team.

CHAPTER 26

WHEN AND WHY THE INCONVENIENT ICE AGE WILL START

All ice ages are inconvenient for the people that have to experience them.

Inconvenient for the plants and animals also.

This chapter covers the important information of when the coming Ice Age will occur and how to get ready for it. Getting ready includes both the national effort and family readiness. As a technical expert in cold weather effect on infrastructure and living things, I'm ready to go with this information and hope it is useful.

WHEN ?

Because Ice Ages are caused by a lowering of the sun's energy, the most accurate way to determine the date is to look at the sun itself.

That's easier said than done because many Climate Change fraud fake scientists comment on this issue. They data is wrong. Either purposely or though lack of understanding.

One who gets it right and who is a world leader in solar science is Dr. Willie Soon. We have mentioned his comments about the fake Climate Change solar data. But we briefly mention them again because it is an important issue. We need the accurate data to know when the Ice Age will begin.

Dr. Soon has commented:

"What can I do to correct these crazy, super wrong errors?"
"Errors in Total Solar Irradiance,"

"The Intergovernmental Panel on Climate Change keeps using the wrong numbers! It's making me feel sick to keep seeing this error. I keep telling them – but they keep ignoring their mistake."

http://www.cfact.org/2018/12/02/dr-willie-soon-versus-the-climate-apocalypse/

So where do we go to get the correct information? Well of course Dr. Soon is one destination. Along with him are several solar scientists who are right on target with their work.

Another is Dr. Valentina Zharkova who made one of the few correct projection on when a previous solar minimum would occur. Her's was one of two models that successfully predicted a small solar minimum previously. Her model was 93% correct regarding all aspects of the solar minimum.

Thus the model we go for help in determining when the Inconvient Ice Age will begin, to is that of Dr. Valintina Zharkova. Professor of Astronomy and Astrophysics at Northumbria University.

She has now predicted, based on a proven accurate model, that we are in for a SUPER Grand Solar Minimum in the near future.

Let's think about this. **Solar minimums mean a cooler earth**, because the sun is cooler.

Grand Solar Minimums mean that the sun will become unusually cool.

And a **SUPER GRAND Solar Minimum means we are in for a very cold, rough ride.** Equivalent to the Little Ice age between 1300 – 1800, or worse.

Professor Zharkova gave a presentation of her **Climate and the Solar Magnetic Field hypothesis** at the Global Warming Policy Foundation in October, 2018. From this presentation, we know the date of the Inconvenient Ice Age.

Zharkova was one of the few that correctly predicted the previous solar cycle (which was #24) would be weaker than cycle 23. 150 models competed to see which would be the most accurate. Professor Zharkova's model was one of only two models that got it right. And secondly, her model was 93% accurate in all aspects of the minimum physics.

Zharkova's models are amazingly accurate. For us most important are that her findings suggest a Super Grand Solar Minimum is going to occur beginning sometime between 2020 and 2035. And that the duration of the solar minimum will be about 350-400 years for the full cycle.

Magnetic field physics, among other matters, confirm her findings. During the previous Little Ice Age, only two magnetic fields of the sun went out of phase.

This time, all four magnetic fields are going out of phase."
https://nextgrandminimum.com/2018/11/22/professor-valentina-zharkova-breaks-her-silence-and-confirms-super-grand-solar-minimum/

NASA CONFIRMS

In addition to this rather scary astrophysics information, the NASA Langley Research Laboratory has also weighed in. The Inconvenitnt Ice Age is at the door step:

Their chief scientist, Dr. Martin Mylnczak, at the NASA Langley Research Center stated in late 2018, that the earth is dumping a frighteningly small amount of heat into space. Meaning we do not have enough heat here on earth. This of course contradicts the Global Warming hysteria of the political Left.

To help track the latest frightening developments, Martin Mlynczak of NASA's Langley Research Center and his colleagues recently introduced the **"Thermosphere Climate Index."**

The Thermosphere Climate Index (TCI) tells how much heat nitric oxide (NO) molecules are dumping into space.

During Solar maximum, when the sun is in a hot cycle, the TCI is high. This means that it will be hot on earth for a long period.

"Right now, it (the TCI) is very low indeed ... 10 times smaller than we see during more active phases of the solar cycle," says Mlynczak

Leading NASA and other Scientists Expect Record Cold could be starting in a matter of months. As we mentioned, Zharkova's solar model has projected that a full fledged ice age will hit sometime around 2035.

A new ice age! Yikes!

"If current trends continue, it could soon set a Space Age record for cold," said Mlynczak late in 2018.

"We're not there quite yet, but it could happen in a matter of months."

OCEANOGRAPHIC DATA TELLS THE SAME STORY

We are going to focus on ocean currents.

Ocean circulation in North Atlantic is at its weakest for 1,500 years - and at levels that previously triggered a mini Ice Age, the study warns

Currents have a 'profound effect' on both North American and European climate

Researchers found a similar weak signal had occurred during part of the previous Little Ice Age. Particularly a cold period which occurred between 1600 and 1850

Researchers studied the Atlantic Meridional Overturning Circulation (AMOC), the branch of the North Atlantic circulation that brings warm surface water toward the Arctic and cold deep water toward the equator.

The research, by Drs. Christelle Not and Benoit Thibodeau from The University of Hong Kong, is interpreted to be a direct consequence of global warming and associated melt of the Greenland Ice-Sheet. But what started out as perhaps another Global Warming support paper, rapidly turned against it.

The research team found a weak signal during a period called the Little Ice Age (a cold spell observed between about 1600 and 1850. The flow rate today is even lower, and sets a record for low intensity flow during the last 1,500 years.

SO WHEN?

NASA Langley Research Laboratory states that it could be any time soon. Ocean currents indicate it could be any time soon. And Dr. Zharkova states it could be any time between now and 2035.

AND WHAT DOES THE START DATE ACTUALLY MEAN IN TERMS OF IT GETTING REALLY REALLY COLD ?

Will IT BECOME COLD SUDDENLY, OR WILL IT TAKE A LONG TIME FOR THE TEMPERATURE TO DROP ?

Unfortunately the cooling effects will apparently occur quite rapidly.

Dr. Jørgen Peder Steffensen is one of the world's most high technology advanced experts on the history of Ice Ages, as determined by ice core sampling. Ice core sampling involves sending a sort of hollow pipe into the frozen arctic ice very deep, then pulling up that ice "core" and running various laboratory analytics on it. This science has in the last several years become extremely accurate.

Steffensen is a Professor of Glaciology at the University of Copenhagen, Denmark, and has published more than 50 papers in his field. He's at the top of his field.

Here is what he has to say about how fast the Ice Age cooling take affects after a Grand Solar minimum is reached:

"Our new, extremely detailed data from the examination of the ice cores shows that in the transition from the ice age to our current warm, interglacial period the climate shift is so sudden that it is as if a button was pressed."

YIKES !

Now that we know it's coming (and let's make it clear, this is based on actual data and not similar to the Global Warming hysterics movement false claims about the end of the world), let's see what are some of the things that can be done to get ready for the Inconvenient Ice Age which is at our door.

And that we will do in the next chapter.

CHAPTER 27

HOW TO GET THE COUNTRY READY... HOW TO GET YOUR FAMILY READY FOR THE SEVERE ICE AGE JUST DOWN THE ROAD

There are two major areas of preparedness:

- The first is national. What we can do to get the infrastructure of the country ready

- The second is personal. What we can do to get our families ready

NATIONAL PREPAREDNESS FOR THE COMING SUPER GRAND INCONVENIENT ICE AGE

President Trump is planning a two billion dollar infrastructure project.

That would be a wonderful opportunity to help get the country ready for long-term very cold temperatures.

The United States, with the exception of a few cold weather states such as North Dakota, is woefully unprepared for an ice age.

If the coming Ice Age begins in 2025, we have just 5 or 6 more years. But it's enough time for a focused well planned program.

If the coming Ice Age begins in 2035, it means we can get totally ready.

So what does "totally ready" mean.

The general areas of US infrastructure readiness include:

- Agriculture
- Transportation
- Water
- Power/ Fuel / Energy (which includes electrical, fossil fuel, nuclear)

In addition to infrastructure, readiness is required in education to help the citizenry train for the coming challenges. As well as training the military, police, and first responders.

AGRICULTURE

Growing seasons are going to be shorter

The temperatures will be lower, and some crops won't be part of the food picture any more

Because transportation will be impacted by the Ice Age, the transporting of food from one area of the country to another will take a hit

So clearly there will be food shortages. During the last Ice Age 1300 to 1800, about 10% of the population died from starvation.

Not only crops, but animals will be impacted. Dairy, poultry will come under extreme pressure, and a major percentage of the animals normally as part of the agricultural system will disappear. Eggs, beef, fish, poultry will be in short supply and certainly not nearly enough for the population.

The food crisis leads to other crises. Malnutrition can lead to illness. It can lead to food riots and violence. To put it bluntly, one neighbor might shoot another to get a loaf of bread. It sounds horrifyingly, but during the last Ice Age violence was rampant.

What small amount of food there is, will face another problem. Cooking will be difficult. Gas and electric cooking possibilities will be severely reduced.

In a sense, moving into a heavy Ice Age, such as we are doing, is like moving back into the stone age.

HOW TO PREPARE THE COUNTRY FOR THE AGRICULTURAL CATASTROPHE

As an engineer specializing in cold weather, I have worked for years with hydroponics systems to raise vegetables, fruits and other plants indoors during the cold winters.

It may be -20 degrees outside, but we are growing massive amounts of tomatoes and lettuce inside happily.

These hydroponic farms in insulated buildings don't need outside sunlight and can grow food no matter what the temperature is outside. Artificial heat and light can do the job, and is doing it every year in northern Scandinavia to add to the summer crops.

In addition to growing food, if the water and power infrastructure is built to take ultra-cold temperatures. As it is in the Scandinavian companies for example, we can make it through a small ice age with less disruption and chaos that will otherwise occur.

We will need the hardened water and power systems to do much of anything during the Ice Age. And that includes growing food indoors to supplement whatever Food we can grow outdoors.

It will be possible to have agricultural activity outdoors, but the results will be less because of colder weather and shorter growing periods. In the case of a strong long Ice Age, most food production would have to move indoors eventually.

Dear reader, I've had the work experience of designing several hydroponic growing centers. These can produce massive amounts of food, no matter what the weather outside. They work. It's a proven technology and we are going to need many things like it in the coming ice age.

Growing periods outside will be shorter, and the temperatures will be colder. Crop growth, food growth will decrease and decrease.

But by getting ready in a massive national and international effort, we can save people and animals from starvation.

Here is what such large hydroponic centers look like:

And what about the cows? Those cute cows that the Global Warming / Climate Change mob wants to kill a lot of. And why, because they produce methane.

In fact, during an ice age, that methane will be a goldmine.

How to we get methane from the cows? Simply by securing their fecal waste, cow dung, and putting it in bio-processing huge centers which will produce methane. That methane can produce heat and also electrical power.

I have designed methane-from-waste generating plants. They work.

And for humanitarian and social justice reasons, we will be saving the cows as well from the genocidal political Leftists.

The only thing standing in our way is the Democrat Party, Social Justice / Democratic Socialist Party.

The struggle is political, judicial, and media. But with the coming Ice Age, comes

the destruction of the lying, cheating, and stealing Democrat Party. And their Global Warming hysteric spawn, the JUSTICE DEMOCRATS / DEMOCRATIC SOCIALISTS.

According to experience in the previous Ice Age, ice ages breed violence.

I suppose the masses of Americans who have been lied to about Global Warming and Climate Change Caused By Human Activity are going to be a bit upset at those who lied. Those who sent the country in the wrong direction, and acted against getting ready for the coming Ice Age.

The Democrat Party may find itself engulfed in a wave of violence. No one wants that. But Ice Ages produce some horrendous results.

TRANSPORTATION

Putting it bluntly, the roads and highways will be filled with snow and ice.

Cars, trucks and buses will be frozen.

Air travel will be difficult if not impossible, runways filled with snow and ice. Aircraft grounded due to ice on their wings and in their engines.

AOC will get her wish about an end to air travel, it will be frozen shut. So will other transportation.

Even horses won't make it. Transportation will consist of what we in science call "Bus # 11". Which means your own two feed.

Scandinavian countries, which are used to dealing with masses of snow in the winter and very cold temperatures, and because they have the necessary equipment already, will make it through. But the US will be frozen shut.

Most of the country will simply be covered with ice and snow in the winter, making it difficult to move.

The snow removal equipment in use in most of the US, is a technology 70 years old, and won't be able to withstand the challenge. Roads will be impassable.

Something like these will be needed, equipment which I work with all the time in brutal cold weather climates where I am doing research.

WATER

Just imagine an ordinary day in New York city. An ordinary winter.

People get up in the morning, and put their glass under the kitchen water XXX, turn the handle and... nothing happens. No water.

Nothing to drink, no way to wash their face, no way to flush the toilet.

And it's not going to be fixed either, because water pipes all over the city have burst from the cold. When water freezes into ice inside water pipes, it expands. The force is enough to burst the pipes like a balloon. There is no hope.

About half of New York city residents are estimated to become victims of the coming Ice Age.

Pipes outside their building, and city water supply systems will simply have soft explosions. The damage will be irreparable. And even if an army of plumbers could replace every pipe, it would just happen again.

And as their buildings get colder from lack of heat, the pipes inside the building will burst as well.

There will be water outside, in the form of ice and snow, but getting to it will be difficult. For a variety of reasons, the elevators in the buildings won't work. And even if people walk down and up their building steps, water from the outside ice and snow will be so contaminated and dirty, they can't use it. Or if they do, they will become sick.

Simply drinking water that is nearly frozen, is enough to bring sickness, in addition to the contamination problems.

As I mentioned, dear readers, I'm an engineer working in alternative energy and in infrastructure design which is meant to with-stand ultra cold temperatures.

There is no university course in this in the US, no college or professional school course. I had to learn it in places which actually use these water infrastructure system that can withstand -50 degrees C (-45 degrees C) with no problem. There I could not only get courses but could participate in the construction of district heating water systems.

Basically the methodology used in Scandinavia, for example, utilizes placement of the pipes as an important measure. It has to be below what is called "the frost line", meaning that below that depth in the ground, things won't freeze.

It's generally about two yards. Thus the pipes are buried much deeper than we generally place them in the US.

In addition, the pipes are insulated. With massive foam insulation which leads to the water pipes become a pipe within a pipe:

Also, in Scandinavian cities, buildings do not produce their own hot water. No boilers inside buildings. The hot water comes from a central city or area water heating plant. Sometimes the heat center is an electrical generating plant, and the waste heat is used to heat water.

This all means that the US is going to have to make a massive shift as to how we transport our water, and heat our water.

No matter what the cost we have to do it. Not doing it means that at least half the residents of large cities will die within weeks or months of the start of the Inconvenient Ice Age.

Imagine life without water being delivered to your home by pipes. This is the Inconvenient Ice Age unless a massive infrastructure effort is made during the next several years.

A major problem with the fake Global Warming / Climate Change Caused by Human Behavior hysteria movements, is that they are sending us in the wrong direction. They are keeping us from preparing for the real emergency.

ELECTICITY, HEATING GAS, GASOLINE, OIL, AND OTHER FOSSIL FUELS

Electricity isn't made by magic. It generally comes from one of two sources: fossil fuels or nuclear power.

Both of these are declared as enemies by the fake Global Warming hysterics. The very things that can save lives during the coming Inconvenient Ice Age, are the things they try out destroy and outlaw.

Even if the Green New Steal fails to destroy fossil fuels, and thereby also destroying electricity, there are going to be fuel and power serious challenges during the coming Inconvenient Ice Age.

One of these problems is transporting the fuels. As mentioned, roads and highways will be filled with ice and snow. So then how to deliver the fuels to power plants that produce electricity and hot water? Actually, there is no way.

No way unless the massive transportation infrastructure project for transportation is carried out. But even with that, fuel deliveries will probably sooner or later grind to a halt.

This is another reason why half the urban population of the country will die in the first few weeks or months. And much more of the population in the first several years of the Inconvenient Ice Age.

SOCIAL COLLAPSE

As the water, food and fuel begins to disappear, we know from the history of the previous Ice Age (1300 – 1800) people turn on each other. That sort of Hollywood future in which barbarians compete for survival will become a reality.

Unless we get ready.

And the Green New Traitors are keeping us from that.

WHAT CAN YOU DO FOR YOUR FAMILY? HOW TO PREPARE?

In the coming Ice Age there will probably be attempted mass migrations. People who manage somehow to survive the first shocks, will want to get to warmer climates.

That is going to be very difficult because after several months or a few years, there won't be any forms of transportation except "Bus Number 11". Meaning our own two legs. Walking.

AOC will get her wish. No cars, no buses, no planes. No nothing. Not even horses because they will have been eaten long before.

Given this bleak future of the coming Inconvenient Ice Age, is there anything we can do for our families ?

Well there is.

First of all, have a plan to get your entire family together. Everyone, aunts, uncles, cousins, everyone. There is strength in numbers. And families won't turn on each other.

Neighbors will, but families won't.

Secondly, have a place. This is of course easier said than done. But getting out of high rise buildings in cities, once electricity and water problems begin, is an absolute necessity. Therefore plan ahead. Try to find family members or friends who have a place outside the cities when the group can stay.

Thirdly, and I hate to mention this, I really do, but whatever your family group has of value, others will try to steal it eventually. It's true that when the going gets tough, the tough get going. But also true that when the going gets touch, many people fall apart and become violent predators. Therefore use the Second Amendment.

In addition, and this is among the most important of realistic preparations: specialized warm clothing, footwear, and tents suitable for cold weather use.

As part of this "warm" category, and among the most important, is that every member of the family have a sleeping bag rated as comfortable to at least -5 to -15 degrees F (- 20 to -26 degrees C).

Understand that other people outside your family will try to steal warm clothing, sleeping bags and tents if they know you have it. The worst of them being former members of the Global Warming hysteria movement, who are in shock because they were so wrong.

Getting out of crowded urban centers is a prerequisite for survival.

Things would not be this tough, were in not for the the hundred billion dollars and many years wasted thus far on non-existent Global Warming.

GETTING ON THE RIGHT TRACK, GETTING ON THE PREPARE THE INFRASTRUCTURE TRACK...

The right track means to prepare infrastructure and our life supports systems of agriculture, power, transportation for freezing weather, that easily can kill half the country and plunge it into chaos and disorder.

So let's fight the good fight. Help the government of President Trump overcome the power of the lie of Global Warming / Climate Change By Human Activity. The false religion of the Democrat Party and its vipers, Justice Democrats / Democrat Socialists or whatever they are called.

The false religion is simply used as a tool by them to control We the People.

And get us ready for what is really at the doorstep. Not hot weather but instead almost endless snow, ice and freezing weather.

Who knows, we might be able to get ready as a country in time. We might be able to largely avoid the collapse and chaos of the society which will occur if we do not prepare for the ice age.

We have technologies, we have unique abilities in the United States. Let's mobilize it to fight against freezing in the coming ice age. It's an uphill battle, but we can do it.

This push at preparation means crushing (politically, financially, and through the courts) many Democrat groups.

One is the Democrat Media which promote the Global Warming lie.

It also means politically crushing the Democrat Party (home to Global Warming policies which are national suicide for the United States) and their voting fraud made of illegal alien voters and counting the votes wrong.

It means outing, and voting out the secret so-called Justice Democrat coup against the United states. Against in fact against liberty and justice.

And in addition relegating to the dust bin of history, those millionaires and billionaires who benefit financially from the Global Warming lies. Or who seek to gain tyrannical power.

One thing to do is to get attorneys who can help, or if you are an attorney, take action. Political, legal, and media action to save your children, save your family. We are heading into freezing hell and not prepared for it because of the Global Warming fraud.

Another is to join and energize the Republican Party. So it becomes the shock troops for liberty and freedom. And crushes evil from the political Left.

Join in. Fight back.

In this political battle, information battle, and legal battle, against the Global Warming hysterics, in all their forms, let's remember the words of David, when he faced Goliath. The Global Warming movement at present has the money and power and a massive blind following. They are Goliath, large, evil, but also vulnerable because their movement and their lives is based on a lie:

Let's keep the words of David in our hearts for this battle. It's a battle we can win.

"You come against me with sword, spear, and javelin, but I come against you in the name of the Lord God Almighty, the God of the Armies of Israel, whom you have defied.

This day the Lord will hand you over to me, and I'll strike you down and cut off your head. Today I will give the carcasses of the Philistine army to the birds of the air and the beasts of the field, and the whole world will know that there is a God in Israel.

All those gathered here will know that it is not by sword or spear that the Lord saves; for the battle is the Lord's and he will give all of you into our hands."

CHAPTER 28

THE SURPRISING NEWS HIDDEN BY THE DEMOCRAT MEDIA COMPLEX:

A REVOLT BY THE PEOPLE AGAINST NATIONAL GOVERNMENTS AND MULTINATIONAL GLOBAL WARMING POLICIES IS TAKING PLACE ACROSS THE WORLD.

EVEN IN SOME STATES IN THE USA IT'S STARTED

LET'S ALL AMERICA REVOLT AGAINST GLOBAL WARMING FRAUD BY USING POLITICAL, JUDICIAL, AND MEDIA EFFORTS

2018 saw the clear beginning revolts by We the People, world wide, against Global Warming tax policies of their governments.

These taxes crush people. They are called "carbon taxes". And the idea behind them is to raise the price of fossil fuels by adding heavy taxes to them. So almost no one can afford to buy them.

Thus, gas for your car, coal or electricity or oil to heat your home / apartment, or hot water for district heating, suddenly costs a lot more.

The idiotic idea of the blind Global Warming politicians that introduce such taxes (such as Trudeau in Canada, and Macron in France), is that if the price goes high enough, people will use less fuel. And that means less CO_2. And that means less Global Warming.

The problems are:

- The amount of CO_2 in the atmosphere, a lot or a little, has zero connection to heating or cooling of the earth surface area

- The taxes make industries and people in the advanced countries which must pay them, non-competitive against other countries that do not have such taxes.

    So jobs and industries where there are Global Warming taxes disappear. The people suffer. But that's all part of the scheme of the wicked secret Leftist political and billionaire forces behind Global Warming fraud. Chaos is their objective. Because on the tailcoats of that chaos, they can ride to power.

    China, India can replace jobs in America or the EU for example. In part because do not have Global Warming policies or such taxes.

EU countries such as Canada, Australia, France and many other EU countries besides France, and several states in the US which have Left wing radical Democrat Party members as governors, do have these crazy carbon taxes. Carbon taxes were an actual government policy for the whole country under Obama. Hundreds of thousands in the energy industries lost their jobs because of this. Electricity prices were planned by Obama to "skyrocket".

Did this bother the Global Warming people? Not one bit. They are all about power.

- People can barely now afford the cost of living. Raising the cost of things which are absolutely necessary, such as fuel for heating, cooking, traveling, means suddenly their financial situation gets a lot worse.

    We the People will not tolerate it. Resist.

There is wisdom in We the People. And in Canada, France, Australia, and some parts of the USA for example, the revolt against the false religion and oppressive destructive taxes of Global Warming has begun.

We now need to carry the political, judicial, and media revolt forward everywhere. Maybe not in the same way as France for example, but at least in politics, the judiciary, and media.

The volcano against the Global Warming and Climate Change lies has been growing and ready to erupt for some time.

One of the first governmental explosions against the false religion of Global Warming took place in June 2018 with the election of Conservative Doug Ford as Premier of Ontario Province in Canada. Ontario is a huge area, about one-fifth the size of the US. And the people voted Global Warming and its policies into the garbage.

In his election campaign, Ford said that his first act as the premier (governor) of the province would be to repeal the Global Warming taxes. To get rid of the carbon tax.

This has of course been totally hidden from the American public by the Democrat Media complex.

Just as the Nazi regime disarmed the people, and kept them ignorant of the truth. So the Democrat Media Complex follows that same script.

In fact they are worse than the NAZI mass media (radio, newspapers, and films at that time). Because the Democrat Media Complex today is doing it happily and voluntarily. No one is forcing them to tell lie, after lie, after lie. But they do it anyway.

The Democrat Media complex is complicit in the crimes of the Global Warming movement.

Now let's get back to the news that the Democrat Media Complex is hiding from the American people:

After Canada, the revolt against CO2 policies and taxes then spread to Australia.

The Prime Minister Malcolm Turnbull was swept into the political trash bin when his Liberal political party tried to pass a carbon tax and other Global Warming punitive legislation against the people.

He failed to get the legislation passed because of opposition of Conservative legislators. And not only did the legislation fail to be approved, Turnbull was replaced as Prime Minister. Another political guillotine against Global Warming fell.

Then in the USA, Democrat and global warming maniac Gov. Jay Inslee of Washington state was the next to taste defeat.

He attempted to pass a measure in the November 2018 elections, which would have placed a heavy carbon tax on the people and businesses of the state.

Although Islee was able to pump $45 million into the campaign, he lost. The state voters rejected the carbon tax. And politically told Islee to "shove it".

The few Republicans in the US Congress who support Global Warming and carbon taxes bit the dust.

In the November 2018 House of Representatives elections, Republican members of the House Climate Solutions Caucus were shown the door in the elections. Good bye.

Among those defeated was caucus co-chair Florida GOP Rep. Carlos Curbelo.

Curbelo proposed a huge carbon tax and was roundly defeated in the election with all of his RINO (Republican in Name Only) henchmen colleagues on the committee who supported the Global Warming legislation.

Then the revolution against Global Warming government policies spread to France.

At the end of 2018, the well-known "Yellow Jacket" revolt caused the government to completely remove the new carbon tax on fuels. And the popularity of the insane Leftists Global Warming believers, specifically the Macon government, plummeted into the gutter.

World-wide, this revolt by the people, as protesters and voters, against the insane Global Warming is happening. It is hidden from the American Public by the Democrat media, which are not media or press, but which are political hacks and mouthpieces for Democrat tyranny. The Democrat Media Complex includes:

**CNN, MSNBC, ABC, CBS, NBC,
The Washington Post, The New York Times,
The Associated Press, Bloomberg News,
National Public Radio (NPR), National Public Broadcasting,
the 95% of the White House Press corps which is radical Left,
Buzzfeed, Politico; Mother Jones, Huff Post and many many additional nests of vipers.**

But the people are awake and are taking steps to stop their governments from committing national suicide. 2018 world-wide has been a bad year for the Global Warming false priests and for their false religion.

We the People can 2019 and 2020 to build the political, judicial, and media barricades against the insanity of Global Warming politicians. And the evil forces behind them, such as Cenk Uygur, Zach Exley, AOC, A Pressley, Rashida Talaib, and many others.

Bigots, racists, antisemites, haters of "white people", haters of Christians:

Do they remind you of anyone? Guess from where the JUSTICE DEMOCRAT inspiration comes:

Exley and Becky Bond are cookie cutter copies of Nazi and Soviet ideology. Down to even using the fascistic art of their heroes.

And Exley and Bond are are in turn puppets of the Leftist billionaire organizations funded mostly by Soros.

Especially now since unhinged Democrat majority in the House or Representatives began on January 3, 2019 there is an increasing movement of open tyranny in American government and politics.

The Democrat Party has a huge majority of 40 seats now in the House of Representatives. For two years. This majority is the result of election fraud, illegal alien voting.

Normally insane asylums are off limits to the media and the public view. But the House of Representatives, with its new Global Warming mob is open for all to see.

The Global Warming policies which they propose have two effects. And make no mistake about it, these destructive effects are intended. Uygur and Exley as well as the dark money billionaire-funded organizations behind them are pushing for it:

First to destroy the US economy by destroying our energy infrastructure.

To eliminate the oil and gas and coal which have in fact become forms of relatively clean energy, and which power our jobs and lives. Without them, collapse and chaos awaits. And collapse and chaos is exactly with the political Left wants. Like Stalin, like Hitler, they can gain power only through collapse of the existing systems. This is one reason they champion open borders.

Second, the insane and fraudulent theories of Global Warming to make it certain that when the New Ice Age hits, the USA will have collapse of the society, food, and energy. Failure to get ready for the coming ice age means death and destruction. It isn't a joke.

When it starts getting colder and colder, really colder, every year, we are going to need oil, gas, and coal more than ever. Without them we will not survive.

If the plans of the Democrat Party Global Warming maniacs are carried out, about 50% of the US population could die rather unpleasant deaths as the New Ice age gains in ferocity. And it is coming sooner than later according to the major scientific markers presented in this book..

The population which survives in the Democrat and Justice Democrat frozen paradise, will actually be in a hell of constant warfare of neighbor against neighbor.

Just imagine, anyone living in a big city. Take for example that paradise of the Left, New York City.

Based on a combination of the cold weather, and of the shortage of fuels caused by the Democrat maniacs, there is going to be a rather unpleasant situation.

The elevators won't be working. The water system will stop functioning because the pipes will burst.

No electricity because the power stations don't have the power to run them. The coal, oil, or gas will be in short supply if the Democrat maniacs have their way. Very few cars, buses, trains or planes.

And as the fuel runs out and people start to freeze, these Leftists who have been lying about fake Global Warming, that it's going to get a lot hotter, will turn on each other.

The solidarity of the tyrannical Left will break down after they start getting cold and hungry.

The Global Warming monsters will kill their former friends, and their neighbors for a blanket or a piece of bread.

Inside every Leftist is a little totalitarian monster, just screaming to get out. The New Ice Age will break down whatever restraints remaining that the Leftists have, and their inner monsters will get out and take control.

In fact, we don't have to wait for the ice age, that is happening right now before our very eyes.

I suppose that like the people of France, Canada, and Australia, who are rising up against the Global Warming fraud, we in the US will wake up also on the national level. We the People will politically, judicially, and in the media crush the Democrat maniacs who are sitting in the driver's seat at the House of Representatives.

We will use political, judicial, and media warfare against them. Protests which lock them down so they can't hurt anyone. Peacefully but with effect.

There is no choice. The New Ice Age, which is coming to your neighborhood very soon, is going to determine matters. And after it begins, it's going to be a matter of survival.

Are we going to have rational and effective policies to meet the challenge? Or are we going to do as the Democrat Party wishes and plunge the nation into chaos, violence, and national suicide?

A NOTE TO ONE OF AMERICA'S LEADING ACTRESSES, AOC

This isn't personal. I just want to mention, what could possibly go wrong by following through on the idea of your puppet masters, THE JUSTICE DEMOCRAT radicals, of eliminating fossil fuel industries and running everything off solar and wind.

Soooo.... what could possibly go wrong?

Let's suppose that the main theme of this book is correct, and a New Ice Age is coming very very soon. Meaning by 2035 at latest, or before.

When the ice age starts getting really cold, the wind turbines will freeze shut. The won't turn. They won't generate any power.

As mentioned previously in this book, I worked and did energy research in Havöysund, Norway, where this freezing of huge wind turbines actually happened. I could walk right up to those huge turbines. Simply because in the winter they did not turn.

And guess what? Even if they did, it takes so much energy and specialized materials to make them, that the production causes about the same CO_2 output as they save in their 25 year working lifespan.

At the end of 25 years, the turbines have to go to the junk yard and be placed by new ones. The idea that alternative energy equipment lasts forever is simply not true. Another Democrat, Justice Democrat fraud.

And when it gets a lot colder, what then will happen regarding the darling of the sneak attack revolutionaries THE JUSTICE DEMOCRATS, photo voltaic electrical generating panels?

When they are covered with snow or ice, they don't work. Power lines will crash to the ground as the result of snow and ice on them. The panels will become useless.

And even if the climate gets very hot, they won't work either. Photo voltaic cells get less and less efficient the hotter it gets. Right now at best they are about 22% efficient. With extreme heat, they will go down to near zero and then burn out after a few years.

But in any case, every 25 years they have to thrown into the garbage because they don't work any more. And they can't be recycled.

What an environmental disaster. Mountains and mountains of solar panels that don't work any more. And the energy needed to recycle even their aluminum frames (a small part of the panel) would be prohibitive. Mass amounts of fossil fuels would be needed to generate the necessary heat just to recycle the aluminum frames..

And finally about photo voltaic electric panels, making them in the first place requires processes that use heat generated by fossil fuels. Yes, those nasty fossil fuels have to be used in huge amounts to make the solar panels. Because we are essentially melting minerals and metals in order to make them.

ALTERNATIVE ENERGY WON'T WORK IN EXTREME HOT OR COLD

OK, so we have seen that if an Ice Age is coming, the alternative energy equipment the Pearl Harbor sneak attackers, the JUSTICE DEMOCRATS, and the whole insane Democrat Party dreams about, won't work.

For more information on the despicable JUSTICE DEMOCRATS see this amazing investigative report on The Reagan at:

- **https://www.youtube.com/watch?v=1h5iv6sECGU**

REMEMBER... if it gets really hot, as the Global Warming religionists chant day and night...

Well unfortunately, solar electric panels won't work, in that heat.

Solar electric panels get less and less efficient the hotter they become. So if we are really going into a time of very hot temperatures, the solar panels will become very inefficient. **Maybe you will need a football field of them to generate enough power to run a toaster.**

As for the wind turbines, look around the world. There are not any in really hot places such as Dubai and the United Arab Emirates. Simply because they can't take extreme heat. Their operating conditions are probably limited to 40 degrees C (104 Fahrenheit). And that's nothing compared to what the Global Warming hysterics are forecasting.

And like the solar electric panels they will become very inefficient and shut down. After 25 years, they too have to be shipped to the garbage. An environmental catastrophe. Imagine endless piles of huge turbines and blades.

Yes, after 25 years of use, all those nice solar electric panels and wind turbines, head to the junk yard. The Lefties don't mention that, and most of them don't even know it.

But our old friends, oil, gas, and coal don't stop working after 25 years. And they not only produce electric power, but produce heat as well.

And when the New Ice Age hits, probably this year, but if not then soon after, we are going to need heat.

And finally regarding the insane theories of the Pearl Harbor sneak attack JUSTICE DEMOCRATS what about those those electric cars and cow farts?

Actually there is no such thing really "electric cars". They are only "coal cars". Or nuclear power cars. These so-called "electric cars" run on coal.

Because the electricity used to charge them comes from coal-powered electrical generating plants.

Coal is the dreaded enemy of the political Left. A target of the Democrat Party. Thus they are targeting their own little darling of "electric cars".

According to the Democrat Party and its fringe crazies, the environmental way to travel is by horse.

But I don't think any of the other Democrat Global Warming people will be use a horse to get home from Washington DC for vacations. It's easy to talk a good game, but to play it is a little more difficult.

But according to the Green New Deal, we the slaves of the tyrannical Left may have to use horses, even for long trips.

Without the aviation fuel, which is a petrochemical and comes from oil which you want to ban, horses are the only alternative if you want to fly somewhere but can't. Or walking might be nice.

But Democrats, you can can forget those nice high tech $200 running shoes. You are going to need big, heavy, warm boots. It's going to get a lot colder.

In the end, AOC and her fellow travelers on the Green New Deal, will be wandering around the countryside with glass or plastic bottle. From the last Coca-Cola or

Pepsi they could get. Those bottles will be used by them to capture... guess what... cow farts.

Yes, the methane from cows can be burned to supply heat. And that heat will be necessary for survival in the coming Ice Age.

Can't we just see AOC, Exley, Uygur, Becky Bond, Maisie Hirono and a few of their billionaire supporters, running after the cows, hoping to get some methane?

CHAPTER 29

THE INCONVENIENT ICE AGE START IS NIPPING AT OUR TOES

When I began this book several months ago, after a few years of research, I thought it would be a light-hearted book. Interesting and funny. Also focused on exposing lies of the political Left and their efforts to grab power based on climate hysteria.

And I hope I achieved that objective to some extent or another.

But the more I wrote and the more new research I did, the more startled I became about a few things:

First, the Global Warming movement is much crazier (from billionaires to the street protesters they pay for), much more greedy, and much more power-hungry than we thought. Here they are:

- The small group of greedy scientists that lie, making huge amount of money for doing it

- And other big money makers, such as the carbon credits merchants, law firms that sue the oil industry, the makers of unnecessary equipment to combat a non-existent Global Warming

- The politicians who use it to whip up the mob, and get votes. All based on a lie

- The lawyers who are making ten, twenty, thirty million dollars at a time in their law suits against oil companies. Or working long term for the radical organizations funded by the Leftist billionaires.

- The dark money billionaires funding the whole thing. Including the Pearl Harbor sneak attack on the free and fair election process... THE JUSTICE DEMOCRATS and the Democrat Party itself.

In 2009 I wrote a paper on Global Warming utilizing chemical analysis of the atmosphere, historical CO2 data, ice at the polar caps data, polar bear data, and most of all solar minimum data (when the sun enters a cooler phase).

And in that paper I estimated that a new ice age could begin in 15 years.

That would be about 2024.

Now it's 2019, and things are already getting a bit unsettling.

Three concurrent solar and global processes are now converging on us:
First, the sun as known by all serious solar scientists is entering a cool phase. Remember Dr. Willy Soon, Astrophysicist. And all the important, legitimate, solar scientists. The cooling cycle of the sun has started has already started. And we are basically in it. And we will arrive at the catastrophic solar minimum very soon, but at latest by 2035.

Second, there is a correlation between ice ages and the ocean current flow rates. The lower the flow rate, the sooner the ice age will be here. We have now entered the lowest current flow rate in 1,500 years. And the same flow rates that existed during the recent mini-Ice age from 1300 to 1800. Some very scary data.

And thirdly, recent data shows that the earth is losing a huge amount of heat daily into the upper atmosphere and stratosphere, into space. This is actually quite frightening.

The Langley Research Center at NASA probably knows more about these heat loss issues than any other single scientific organization.

Their chief scientist, chief researcher, Dr. Martin Mylnczak, at the NASA Langley Research Center stated in late 2018, that the ice age is basically on the doorstep. And could be here in a few months.

The catastrophic solar minimum level may not be reached until later, possibly as late as 2035, but we are on the way down, and could possibly go over the cliff at any moment.

This is the opinion of all key international and US solar, climate scientists. Not only in the US, but in the EU, Canada, Australia, Russia, Japan and so on.

The information and news about it, simply isn't making it into the mainstream media, and particularly not making it into the Democrat Party Media Complex which is pushing the false Global Warming & Climate Change from human activity religion.

The role of a free press is actually to inform the citizenry. But the Democrat Media Complex is not doing that. And is actually destroying freedom of the press by trying to crush every idea they wish to suppress. Always using lies and omitting the truth.

And it's not just CNN and MSNBC. It's ABC, CBS, NBC, The New York Times, The Washington Post, most of the White House Press Corps, The Associated Press, NPR, along with their internet partners in crime, Buzzfeed, Politico, HuffPost and so on.

WHAT ABOUT THE ICE AGE ITSELF

Regarding the new ice age itself, we can only hope that this is going to me a "small ice age", and maybe last from 50 to 500 years and not be so cold that all human and animal life on earth will die.

If it's a full ice age, then it's eventually the end. Of everything.

But in any case, a little ice age, or a major one, we are the ones who are going to decide how we go.

If governments actually get ready, as much as they can, for an ice age, we can avoid some of the most awful side effects. These side effects include starvation of a significant part of the population, a collapse of law and order.

During an ice age, organized and random killings are going to become commonplace and on a large-scale if we do not get ready.

Sex crimes rampant.

Starvation and death due to cold, rampant.

The Global Warming /Climate Change From Human Activity mob, instead of saving the planet are destroying it. They and their dark political allies such as JUSTICE DEMOCRATS and the Leftist billionaires financing them, are keeping us from getting ready.

I suppose that one of the things that could happen in an ice age is this. As the society collapses, when food deliveries stop, when electricity is no longer available, there will be new Salem Witch Trials.

The "Witches" are going to be the Global Warming gurus, billionaire financiers, politicians, lawyers, street protesters. And the Pearl Harbor sneak attack on democracy gang, the JUSTICE DEMOCRATS.

And a society in collapse may find that the best way to deal with them is simply to push them into the great outdoors they hysterically proclaimed in superiority would become very hot. And let them become big cubes of ice.

It's not too late to start preparing, and avoid the worst side effects. To avoid a total collapse of law and order. And of the systems we need to survive, such as electricity and food and transportation and water. Yes water.

When we're in the ice age, pipes, both in private dwellings and the carrier pipes of local government are going to burst. Both water pipes and sewerage pipes.

They are not made, or put in the right places, for this kind of weather. Buckets will become the new toilets and water transportation system in Ice Age America.

What this means with the water pipes is that although we will be surrounded by ice and snow, which are made of water, we will die from lack of water.

Both because the pipes will freeze and also eventually we will have no way to melt the snow and ice. Of course, in emergencies it's possible just to shovel the snow and ice into our mouths. But after a while, people can't take the additional cold. Cold from outside and cold from within the human body at the same time, is very difficult to withstand.

When our bodies are cold, this eating of ice becomes a difficult thing to do.

And in any case, in cities the snow and ice will quickly become too dirty and contaminated to drink. In large urban centers, that is going to happen very fast.

President Donald J. Trump is absolutely correct about Global Warming, and he is one of the very few people in government standing between the US and a frozen collapse.

Take political action to support those who are fighting the Global Warming lies. We are not helpless. Not hopeless. If it's a "small ice age" about 50% of the US population could make it through if infrastructure is prepared. And law and order matters readied for the disaster.

Flood the Republican party with enthusiastic political shock troops to support the president.

It's interesting, that underdeveloped agrarian societies, without elevators, without water delivery methods, and without heating systems other than burning natural fuels, could fare much better than developed countries when a new ice age hits.

But we are a developed country, dependent on energy and transportation. We need to be calm and get ready by hardening our energy, food, transportation, and water infrastructure against freezing. Now.

That's better than nothing, so let's get started and do the best we can.

To end this chapter on a happy note, here is a list of **a few of the world leading scientists who at the beginning of the Global Warming fraud, saw the truth and warned the United Nations that Global Warming is a fake.**

Presenting the entire list of over 100,000 scientists who have recognized that Global Warming is a fraud would take too much paper. Although the paper may come in handy later as fuel.

But reading 100,000 names would be time consuming. So we just list some of the pioneer scientists who exposed the fraud in addition to the other heroic scientists against the Global Warming lie listed in this paper.

This list is just one climate organization among many, which understands Global Warming / Climate change is a fraud.

The list contains Americans and international scientists, the absolute best world wide professors and private industry scientists. They have one clear message: man made Global Warming is a fraud.

The list is impressive not only because it is a long list, but because the people in it are amazing. Try to read the whole list if you can. Maybe not in one sitting. But as and after you read, remember they all agree... Global Warming is a massive scientific and financial fraud:

1. Habibullo I. Abdussamatov, Dr. Sci., mathematician and astrophysicist, Head of the Russian-Ukrainian Astrometria project on the board of the Russian segment of the ISS, Head of Space Research Laboratory at the Pulkovo Observatory of the Russian Academy of Sciences

2. Göran Ahlgren, docent organisk kemi, general secretary of the Stockholm Initiative, Professor of Organic Chemistry

3. Syun-Ichi Akasofu, PhD, Professor of Physics, Emeritus and Founding Director, International Arctic Research Center of the University of Alaska

4. J.R. Alexander, Professor Emeritus, Dept. of Civil Engineering, University of Pretoria, South Africa; Member, UN Scientific and Technical Committee on Natural Disasters, 1994-2000

5. Jock Allison, PhD, ONZM, formerly Ministry of Agriculture Regional Research Director,

6. Bjarne Andresen, PhD, dr. scient, physicist, published and presents on the impossibility of a "global temperature", Professor, The Niels Bohr Institute, University of Copenhagen

7. Timothy F. Ball, PhD, environmental consultant and former climatology professor, University of Winnipeg, Member, Science Advisory Board, ICSC,

8. Douglas W. Barr, BS (Meteorology, University of Chicago), BS and MS (Civil Engineering, University of Minnesota), Barr Engineering Co. (environmental issues and water resources)

9. Romuald Bartnik, PhD (Organic Chemistry), Professor Emeritus, Former chairman of the Department of Organic and Applied Chemistry, climate work in cooperation with Department of Hydrology and Geological Museum, University of Lodz

10. Colin Barton, B.Sc., PhD, Earth Science, Principal research scientist (retd), Commonwealth Scientific and Industrial Research Organisation (CSIRO)

11. Joe Bastardi, BSc, (Meteorology, Pennsylvania State), meteorologist, State College

12. Ernst-Georg Beck, Dipl. Biol. (University of Freiburg), Biologist

13. David Bellamy, OBE, English botanist, author, broadcaster, environmental campaigner, Hon. Professor of Botany (Geography), University of Nottingham, Hon. Prof. Faculty of Engineering and Physical Systems, Central Queensland University, Hon. Prof. of Adult and Continuing Education, University of Durham, United Nations Environment Program Global 500 Award Winner, Dutch Order of The Golden Ark

14. M. I. Bhat, Professor & Head, Department of Geology & Geophysics, University of Kashmir

15. Ian R. Bock, BSc, PhD, DSc, Biological sciences (retired

16. Sonja A. Boehmer-Christiansen, PhD, Reader Emeritus, Dept. of Geography, Hull University, Editor - Energy&Environment, Multi-Science

17. Atholl Sutherland Becky Bond, PhD (Geology, Princeton University), Regional Geology, Tectonics and Mineral Deposits

18. Stephen C. Becky Bond, PhD (Environmental Science, State University of New York), District Agriculture Agent, Assistant Professor, University of Alaska Fairbanks, Ground Penetrating Radar Glacier research

19. James Buckee, D.Phil. (Oxon), focus on stellar atmospheres

20. Dan Carruthers, M.Sc., Arctic Animal Behavioural Ecologist, wildlife biology consultant specializing in animal ecology in Arctic and Subarctic regions

21. Robert M. Carter, PhD, Professor, Marine Geophysical Laboratory, James Cook University

22. Dr. Arthur V. Chadwick, PhD, Geologist, dendrochronology (analyzing tree rings to determine past climate) lecturing, Southwestern Adventist University

23. George V. Chilingar, PhD, Member, Russian Academy of Sciences, Moscow President, Russian Academy of Natural Sciences, U.S.A. Section, Emeritus Professor

of Civil and Environmental Engineering, University of Southern California

24. Ian D. Clark, PhD, Professor (isotope hydrogeology and paleoclimatology), Dept. of Earth Sciences, University of Ottawa

25. Charles A. Clough, BS (Mathematics, Massachusetts Institute of Technology), MS (Atmospheric Science, Texas Tech University), former (to 2006) Chief of the US Army Atmospheric Effects Team at Aberdeen Proving Ground, Maryland

26. Paul Copper, BSc, MSc, PhD, DIC, FRSC, Professor Emeritus, Department of Earth Sciences, Laurentian University

27. Piers Corbyn, MSc (Physics (Imperial College London)), ARCS, FRAS, FRMetS, astrophysicist (Queen Mary College, London), consultant, founder WeatherAction long range forecasters

28. Allan Cortese, meteorological researcher and spotter for the National Weather Service, retired computer professional

29. Richard S. Courtney, PhD, energy and environmental consultant, IPCC expert reviewer

30. Susan Crockford, PhD (Zoology/Evolutionary Biology/Archaeozoology), Adjunct Professor (Anthropology/Faculty of Graduate Studies), University of Victoria

31. (Claude Culross, PhD (Organic Chemistry

32. Joseph D'Aleo, BS, MS (Meteorology, University of Wisconsin), Doctoral Studies (NYU), Executive Director - ICECAP (International Climate and Environmental Change Assessment Project), Fellow of the AMS, College Professor Climatology/Meteorology, First Director of Meteorology The Weather Channel

33. Chris R. de Freitas, PhD, Climate Scientist, School of Environment, The University of Auckland

34. Willem de Lange, MSc (Hons), DPhil (Computer and Earth Sciences), Senior Lecturer in Earth and Ocean Sciences, Waikato University

35. James DeMeo, PhD (University of Kansas 1986, Earth/Climate Science), now in Private Research,

36. David Deming, PhD (Geophysics), Associate Professor, College of Arts and Sciences, University of Oklahoma

37. James E Dent; B.Sc., FCIWEM, C.Met, FRMetS, C.Env., Independent Consultant, Member of WMO OPACHE Group on Flood Warning

38. Robert W. Durrenberger, PhD, former Arizona State Climatologist and President of the American Association of State Climatologists, Professor Emeritus of Geography, Arizona State University

39. Don J. Easterbrook, PhD, Emeritus Professor of Geology, Western Washington, University

40. Per Engene, MSc, Biologist, Bø i Telemark, Norway, Co-author The Climate. Science and Politics (2009)

41. Robert H. Essenhigh, PhD, E.G. Bailey Professor of Energy Conversion, Dept. of Mechanical Engineering, The Ohio State University

42. David Evans, PhD (EE), MSc (Stat), MSc (EE), MA (Math), BE (EE), BSc, mathematician, carbon accountant and modeler, computer and electrical engineer and head of 'Science Speak', Scientific Advisory Panel member - Australian Climate Science Coalition

43. Sören Floderus, PhD (Physical Geography (Uppsala University)), coastal-environment specialization

44. Louis Fowler, BS (Mathematics), MA (Physics), 33 years in environmental measurements (Ambient Air Quality Measurements)

45. Stewart Franks, PhD, Professor, Hydroclimatologist, University of Newcastle

46. Gordon Fulks, PhD (Physics, University of Chicago), cosmic radiation, solar wind, electromagnetic and geophysical phenomena

47. R. W. Gauldie, PhD, Research Professor, Hawai'i Institute of Geophysics and Planetology, School of Ocean Earth Sciences and Technology, University of Hawai'i

48. David G. Gee, Professor of Geology (Emeritus), Department of Earth Sciences, Uppsala University

49. Lee C. Gerhard, PhD, Senior Scientist Emeritus, University of Kansas, past director and state geologist, Kansas Geological Survey

50. Gerhard Gerlich, Dr.rer.nat. (Mathematical Physics: Magnetohydrodynamics) habil. (Real Measure Manifolds), Professor, Institut für Mathematische Physik, Technische Universität Carolo-Wilhelmina zu Braunschweig, Braunschweig, Germany, Co-author of "Falsification Of The Atmospheric CO2 Greenhouse Effects Within The Frame Of Physics", Int.J.Mod.Phys.,2009

51. Albrecht Glatzle, PhD, ScAgr, Agro-Biologist and Gerente ejecutivo, Tropical pasture research and land use management, Director científico de INTTAS,

52. Fred Goldberg, PhD, Adj Professor, Royal Institute of Technology (Mech, Eng.), Secretary General KTH International Climate Seminar 2006 and Climate analyst and member of NIPCC

53. Wayne Goodfellow, PhD (Earth Science), Ocean Evolution, Paleoenvironments, Adjunct Professor, Senior Research Scientist, University of Ottawa, Geological Survey of Canada

54. Thomas B. Gray, MS, Meteorology, Retired, USAF.

55. Vincent Gray, PhD, New Zealand Climate Coalition, expert reviewer for the IPCC, author of The Greenhouse Delusion: A Critique of Climate Change 2001

56. William M. Gray, PhD, Professor Emeritus, Dept. of Atmospheric Science, Colorado State University, Head of the Tropical Meteorology Project

57. Kenneth P. Green, M.Sc. (Biology, University of San Diego) and a Doctorate in Environmental Science and Engineering from the University of California at Los Angeles, Resident Scholar, American Enterprise Institute

58. Charles B. Hammons, PhD (Applied Mathematics), systems/software engineering, modeling & simulation, design, Consultant

59. William Happer, PhD, Cyrus Fogg Bracket Professor of Physics (research focus is interaction of light and matter, a key mechanism for global warming and cooling), Princeton University; Former Director, Office of Energy Research (now Office of Science), US Department of Energy (supervised climate change research), Member - National Academy of Sciences of the USA, American Academy of Arts and Sciences, American Philosophical Society

60. Howard Hayden, PhD, Emeritus Professor (Physics), University of Connecticut, The Energy Advocate

61. Ross Hays, Atmospheric Scientist, NASA Columbia Scientific Balloon Facility

62. James A. Heimbach, Jr., BA Physics (Franklin and Marshall College), Master's and PhD in Meteorology (Oklahoma University), Prof. Emeritus of Atmospheric Sciences (University of North Carolina)

63. Ole Humlum, PhD, Professor, Department of Physical Geography, Institute of Geosciences, University of Oslo, Oslo, Norway

64. Craig D. Idso, PhD, Chairman of the Board of Directors of the Center for the Study of Carbon Dioxide and Global Change

65. Sherwood B. Idso, PhD, President, Center for the Study of Carbon Dioxide and Global Change

66. Terri Jackson, MSc MPhil., Director, Independent Climate Research Group, Northern Ireland and London (Founder of the Energy Group at the Institute of Physics, London), U.K.

67. Albert F. Jacobs, Geol.Drs., P. Geol., Calgary, Alberta, Canada

68. Zbigniew Jaworowski, PhD, DSc, professor of natural sciences, Senior Science Adviser of Central Laboratory for Radiological Protection, researcher on ice core CO_2 records

69. Terrell Johnson, B.S. (Zoology), M.S. (Wildlife & Range Resources, Air & Water Quality), Principal Environmental Engineer, Certified Wildlife Biologist

70. Bill Kappel, BS (Physical Science-Geology), BS (Meteorology), Storm Analysis, Climatology, Operation Forecasting, Vice President/Senior Meteorologist, Applied Weather Associates, LLC, University of Colorado

71. Wibjörn Karlén, MSc (quaternary sciences), PhD (physical geography), Professor emeritus, Stockholm University, Department of Social and Economic Geography, Geografiska Annaler Ser. A,

72. Olavi Kärner, Ph.D., Extraordinary Research Associate; Dept. of Atmospheric Physics, Tartu Observatory

73. David Kear, PhD, FRSNZ, CMG, geologist, former Director-General of NZ Dept. of Scientific & Industrial Research, Whakatane

74. Madhav L. Khandekar, PhD, consultant meteorologist, (former) Research Scientist, Environment Canada, Editor "Climate Research" (03-05), Editorial Board Member "Natural Hazards, IPCC Expert Reviewer 2007

75. Leonid F. Khilyuk, PhD, Science Secretary, Russian Academy of Natural Sciences, Professor of Engineering, University of Southern California

76. William Kininmonth MSc, MAdmin, former head of Australia's National Climate Centre and a consultant to the World Meteorological organization's Commission for Climatology

77. Gary Kubat, BS (Atmospheric Science), MS (Atmospheric Science), professional meteorologist last 18 years

78. Roar Larsen, Dr.ing.(PhD), Chief Scientist, SINTEF (Trondheim, Norway), Adjunct Professor, Norwegian University of Science and Technology

79. Douglas Leahey, PhD, meteorologist and air-quality consultant, President - Friends of Science,

80. Jay Lehr, BEng (Princeton), PhD (environmental science and ground water hydrology), Science Director, The Heartland Institute.

81. Edward Liebsch, BS (Earth Science & Chemistry), MS (Meteorology, Pennsylvania State University), Senior Air Quality Scientist, HDR Inc

82. Dr. Richard S. Lindzen, Alfred P. Sloan professor of meteorology, Dept. of Earth, Atmospheric and Planetary Sciences, Massachusetts Institute of Technology

83. Peter Link, BS, MS, PhD (Geology, Climatology), Geol/Paleoclimatology, retired, Active in Geol-paleoclimatology, Tulsa University and Industry

84. Anthony R. Lupo, Ph.D., Professor of Atmospheric Science, Department of Soil, Environmental, and Atmospheric Science, University of Missouri

85. Horst Malberg, PhD, former director of Institute of Meteorology, Free University of Berlin

86. Björn Malmgren, PhD, Professor Emeritus in Marine Geology, Paleoclimate Science, Goteborg University

87. Fred Michel, PhD, Director, Institute of Environmental Sciences, Associate Professor of Earth Sciences, Carleton University

88. Ferenc Mark Miskolczi, PhD, atmospheric physicist, formerly of NASA's Langley Research Center

89. Asmunn Moene, PhD, MSc (Meteorology), former head of the Forecasting Centre, Meteorological Institute

90. Cdr. M. R. Morgan, PhD, FRMetS, climate consultant, former Director in marine meteorology policy and planning in DND Canada, NATO and World Meteorological Organization and later a research scientist in global climatology at Exeter University, UK

91. Nils-Axel Mörner, PhD (Sea Level Changes and Climate), Emeritus Professor of Paleogeophysics & Geodynamics, Stockholm University

92. Robert Neff, M.S. (Meteorology, St Louis University), Weather Officer, USAF; Contractor support to NASA Meteorology Satellites

93. John Nicol, PhD, Physics, (Retired) James Cook University, Chairman - Australian Climate Science Coalition

94. Ingemar Nordin, PhD, professor in philosophy of science (including a focus on "Climate research, philosophical and sociological aspects of a politicised research area"), Linköpings University

95. David Nowell, M.Sc., Fellow of the Royal Meteorological Society, former chairman of the NATO Meteorological Group

96. James J. O'Brien, PhD, Emeritus Professor, Meteorology and Oceanography, Florida State University

97. Peter Oliver, BSc (Geology), BSc (Hons, Geochemistry & Geophysics), MSc (Geochemistry), PhD (Geology), specialized in NZ quaternary glaciations, Geochemistry and Paleomagnetism, previously research scientist for the NZ Department of Scientific and Industrial Research

98. Cliff Ollier, D.Sc., Professor Emeritus (School of Earth and Environment), Research Fellow, University of Western Australia

99. Garth W. Paltridge, BSc Hons (Qld), MSc, PhD (Melb), DSc (Qld), Emeritus Professor, Honorary Research Fellow and former Director of the Institute of Antarctic and Southern Ocean Studies, University of Tasmania, Hobart, Visiting Fellow, RSBS, ANU, Canberra, ACT

100. R. Timothy Patterson, PhD, Professor, Dept. of Earth Sciences (paleoclimatology), Carleton University, Chair - International Climate Science Coalition

101. Alfred H. Pekarek, PhD, Associate Professor of Geology, Earth and Atmospheric Sciences Department, St. Cloud State University

102. Ian Plimer, PhD, Professor of Mining Geology, The University of Adelaide; Emeritus Professor of Earth Sciences, The University of Melbourne

103. Daniel Joseph Pounder, BS (Meteorology, University of Oklahoma), MS (Atmospheric Sciences, University of Illinois, Urbana-Champaign); Weather Forecasting, Meteorologist, WILL AM/FM/TV, the public broadcasting station of the University of Illinois

104. Brian Pratt, PhD, Professor of Geology (Sedimentology), University of Saskatchewan

105. Harry N.A. Priem, PhD, Professor (retired) Utrecht University, isotope and planetary geology, Past-President Royal Netherlands Society of Geology and Mining, former President of the Royal Geological and Mining Society of the Netherlands

106. Tom Quirk, MSc (Melbourne), D Phil, MA (Oxford), SMP (Harvard), Member of the Scientific Advisory Panel of the Australian Climate Science Coalition, Member Board Institute of Public Affairs

107. George A. Reilly, PhD (Geology)

108. Robert G. Roper, PhD, DSc (University of Adelaide, South Australia), Emeritus Professor of Atmospheric Sciences, Georgia Institute of Technology

109. Arthur Rorsch, PhD, Emeritus Professor, Molecular Genetics, Leiden University, retired member board Netherlands Organization Applied Research TNO

110. Curt Rose, BA, MA (University of Western Ontario), MA, PhD (Clark University), Professor Emeritus, Department of Environmental Studies and Geography, Bishop's University
111. Rob Scagel, MSc (forest microclimate specialist), Principal Consultant - Pacific Phytometric Consultants

112. Clive Schaupmeyer, B.Sc., M.Sc., Professional Agrologist (awarded an Alberta "Distinguished Agrologist"), 40 years of weather and climate studies with respect to crops

113. Bruce Schwoegler, BS (Meteorology and Naval Science, University of Wisconsin-Madison), Chief Technology Officer, MySky Communications Inc, meteorologist,

science writer and principal/co-founder of MySky

114. John Shade, BS (Physics), MS (Atmospheric Physics), MS (Applied Statistics), Industrial Statistics Consultant, GDP

115. Gary Sharp, PhD, Center for Climate/Ocean Resources Study

116. Thomas P. Sheahen, PhD (Physics, Massachusetts Institute of Technology), specialist in renewable energy, research and publication (Applied Optics) in modeling and measurement of absorption of infrared radiation by atmospheric CO_2.

117. Paavo Siitam, M.Sc., agronomist and chemist

118. L. Graham Smith, PhD, Associate Professor of Geography, specialising in Resource Management, University of Western Ontario
119.
120. Roy W. Spencer, PhD, climatologist, Principal Research Scientist, Earth System Science Center, The University of Alabama
121. Walter Starck, PhD (Biological Oceanography), marine biologist (specialization in coral reefs and fisheries), author, photographer,

122. Peter Stilbs, TeknD, Professor of Physical Chemistry, Research Leader, School of Chemical Science and Engineering, Royal Institute of Technology (KTH), member of American Chemical Society and life member of American Physical Society, Chair of "Global Warming - Scientific Controversies in Climate Variability", International seminar meeting at KTH, 2006

123. Arlin Super, PhD (Meteorology), former Professor of Meteorology at Montana State University, retired Research Meteorologist, U.S. Bureau of Reclamation,

124. George H. Taylor, B.A. (Mathematics, U.C. Santa Barbara), M.S. (Meteorology, University of Utah), Certified Consulting Meteorologist, Applied Climate Services, LLC, Former State Climatologist (Oregon), President, American Association of State Climatologists (1998-2000)

125. Mitchell Taylor, PhD, Biologist (Polar Bear Specialist), Wildlife Research Section, Department of Environment,

126. Hendrik Tennekes, PhD, former director of research, Royal Netherlands Meteorological Institute

127. Frank Tipler, PhD, Professor of Mathematical Physics, astrophysics, Tulane University

128. Edward M. Tomlinson, MS (Meteorology), Ph.D. (Meteorology, University of Utah), President, Applied Weather Associates, LLC (leader in extreme rainfall storm analyses), 21 years US Air Force in meteorology (Air Weather Service)

129. Ralf D. Tscheuschner, Dr.rer.nat. (Theoretical physics: Quantum Theory), Freelance Lecturer and Researcher in Physics and Applied Informatics, Hamburg,

Germany. Co-author of "Falsification of The Atmospheric CO2 Greenhouse Effects Within The Frame Of Physics, Int.J.Mod.Phys. 2009

130. Gerrit J. van der Lingen, PhD (Utrecht University), geologist and paleoclimatologist, climate change consultant, Geoscience Research and Investigations

131. A.J. (Tom) van Loon, PhD, Professor of Geology (Quaternary Geology), Adam Mickiewicz University, Poznan, Poland; former President of the European Association of Science Editors

132. Gösta Walin, PhD in Theoretical physics, Professor emeritus in oceanography, Earth Science Center, Göteborg University

133. Neil Waterhouse, PhD (Physics, Thermal, Precise Temperature Measurement), retired, National Research Council, Bell Northern Research,

134. Anthony Watts, 25-year broadcast meteorology veteran and currently chief meteorologist for KPAY-AM radio. In 1987, he founded ItWorks, which supplies custom weather stations, Internet servers, weather graphics content, and broadcast video equipment. In 2007, Watts founded SurfaceStations.org, a Web site devoted to photographing and documenting the quality of weather stations across the U.S.,

135. Charles L. Wax, PhD (physical geography: climatology, LSU), State Climatologist – Mississippi, past President of the American Association of State Climatologists, Professor, Department of Geosciences

136. James Weeg, BS (Geology), MS (Environmental Science), Professional Geologist/hydrologist, Advent Environmental Inc

137. Forese-Carlo Wezel, PhD, Emeritus Professor of Stratigraphy (global and Mediterranean geology, mass biotic extinctions and paleoclimatology), University of Urbino
138.
139. Boris Winterhalter, PhD, senior marine researcher (retired), Geological Survey of Finland, former adjunct professor in marine geology,

140. David E. Wojick, PhD, PE, energy and environmental consultant, Technical Advisory Board member - Climate Science Coalition of America

141. Raphael Wust, PhD, Adj Sen. Lecturer, Marine Geology/Sedimentology, James Cook University,

142. Stan Zlochen, BS (Atmospheric Science), MS (Atmospheric Science), USAF (retired),

143. Dr. Bob Zybach, PhD (Oregon State University (OSU), Environmental Sciences Program), MAIS (OSU, Forest Ecology, Cultural Anthropology, Historical Archaeology), BS (OSU College of Forestry), President, NW Maps Co., Program Manager, Oregon Websites and Watersheds Project, Inc.,

The above list is from the Copenhagen Climate Challenge. And is just a tiny fraction of actual scientists who recognize the Global Warming and Climate Changed Caused by Human Activity are frauds.

May America be blessed, a country with the skills and courage that can help get ready for the coming Ice Age, if the people, private citizens industry, and the government take action.

It's up to us as We the People to act politically, through the media, and through the courts against the Global Warming fraud and the JUSTICE DEMOCRATS / DEMOCRATIC SOCIALISTS.

Everyone can help. This book is a call to political, media, and legal arms to save as many families as we can from the coming new Ice Age by getting the country ready.

CHAPTER 30

THIS IS A BONUS CHAPTER FOR THE READERS OF THIS BOOK.

IT'S ABOUT THE NEW GLOBAL WARMING BOARD GAME.

GLOBAL WARMING
FOR FUN AND PROFIT

You can find it at your local department store or on line:

After having in the last chapter, number 22, delved into the Nazi roots and methodology of the Global Warming movement. And reading about it's genocidal tendencies, and the Nazi and Soviet propaganda methods of the DEMOCRAT PARTY, JUSTICE DEMOCRATS / DEMOCRATIC SOCIALISTS...

it might now be time to have a little fun and relax a bit before heading out to political, judicial, and media battle.

What better way than playing a board game with family or friends.

This chapter is about a wonderful new game, based on Global Warming fraud. If nothing else in this book has convinced our dear readers who are from the political

Left that Global Warming / Climate Change From Human activity is a hoax, then perhaps this little chapter will do it.

Republicans and Conservatives already knew and know that Global Warming is a hoax. But simply playing the game is tons of fun, and this game will be fun for everyone.

After the New Ice Age begins, and let's hope it's a Small Ice Age just a few hundred years, not a big severe one... things will change. And we are going to have to find a way to be happy at home with our families without a lot of what we have today. Doing without the usual electronics because there will be no electricity for half of the country or more. Perhaps all of it eventually.

Slowly but surely things we take for granted will conk out in an ice age. Electrical power will be one of them. It won't happen immediately. But as it gets colder and colder for longer and longer, eventually it will. Good bye electricity.

No more computers, video games, smart phones.

Therefore it is a pleasure for our publishing company, in addition to this book, to manufacture something that can help you have some fun in the freezing cold.

Before video games existed, the big game sensations were board games.

These consist, for those of you who have not played them, of a strong paper board about 30" X 30" square.

Here is a photo of one of them which is very famous and a lot of fun.

It doesn't need electricity, and it's great for family members of all ages.

GLOBAL WARMING FOR FUN AND PROFIT is just as much fun. And because it doesn't need electricity, it's perfect for use during the coming ice age.

Examples of board games we can buy today are something like: Monopoly, Checkers, Chess, and so on. Now we are adding a new one.

The name of this new game is, as you know...

GLOBAL WARMING FOR FUN AND PROFIT.

Unlike other board games in which the players go either around and around a track on the outside edge area of the board with little plastic or metal pieces which represent them... or sometimes go back and forth with the little pieces of the game...

GLOBAL WARMING FOR FUN AND PROFIT has a track that goes only in one direction.

And that direction is a false direction, a false path created by the Global Warming / Climate Change movement. That false path takes each player crashing off the edge of the table. And plummeting not into a hot area, but into a frozen wilderness.

WHO WINS THE GAME ?

The one who stays on the table with their little game piece, after all the other player's pieces fall off into the ice, wins.

Note that the game does not come with a supply of ice. But that's no problem since every single family is going to have more ice than they know what to do with.

When the pieces of the losing players fall of the table, pick them up quickly so they don't get stuck on the ice covering the floor of your home. By the time you play this game, the New Ice Age will have begun and at least a small layer of frost or ice will cover almost everything in your home that doesn't move.

In this wonderful board game, players move forward on the fake Global Warming highway by rolling dice. It means they can move forward from 2 to 12 squares after each roll of the two dice.

In most board games, it's good to get big numbers on your roll of the dice, but not in GLOBAL WARMING FOR FUN AND PROFIT.

The longer you stay on the table and do not fall off the road of squares at the end of the table, the better your chance to win.

There are 75 squares and after that a player's little piece falls of the table onto the bucket of ice from outside you place on the floor. After the piece of a player falls off into the ice, they are then out of the game.

However, the game is even more exciting than that. If such a thing could be possible...

Each square in GLOBAL WARMING has something written on it!

For example if you roll a 5 with the dice. Square number 5 says:

"YOUR HOME WATER SYSTEM FREEZES, MOVE FORWARD THREE STEPS TOWARDS THE END"

Or if you roll a 7 with the dice, square 7 says:

"TRUCKS AND CARS FREEZE, MOTOR TRANSPORTATION STOPS, MOVE FORWARD 3 STEPS TOWARDS THE END"

If a player rolls a 2, they move forward to square 2, which says:

"TAKE A QUESTION CARD"

The game comes with a small deck of question cards. The cards have questions which are all related to the fact that Global Warming never happened, and that we are starting to freeze in the New Ice Age.

Board games are fun, as you can see from these photos of the wonderful Parker Brothers game, Monopoly. Note that Monopoly and GLOBAL WARMING FOR FUN AND PROFIT are not related in any way. Except that they are both tons of fun.

Examples of questions are (we don't want to give too many away in this article and dampen the thrill of playing the game itself)...

Who told the most lies about the fake Global Warming?

* Al Gore
* Tom Steyer
* Barak Obama
* CNN
* Michael Bloomberg
* Nancy Pelosi and Bill Nye holding hands together
* AOC
* The cows
* All of the above

Another example of questions from the GLOBAL WARMING BOARD GAME is:

Right now you are very cold. What can you do about it?

* Go outside and get even colder
* Burn all your Democrat books and pamphlets to keep warm
* Throw ice balls at your local Global Warming politicians
* Take your old Beto posters, and burn them
* Join the Republican Party
* Apologize to your parents and friends who warned you

Don't you wish you had realized the truth, that Global Warming is a fraud? So you could have campaigned and protested in order to get American ready for the Ice Age?

* Yes
* Absolutely
* 100%

* Sure, I regret what I did
* No, I'm really happy to freeze and lose everything. I'm a mind-numbed, brainwashed person

Don't you wish you had voted Republican? And seen through the Lies of the Democrat Party and JUSTICE DEMOCRATS?

* Yes
* Absolutely
* 100%
* I'm so upset now, when I get the chance I will vote for Republicans, twice or three times in the same election. Just like I did for the Democrats

This GLOBAL WARMING fun-for the-whole-family game can be pre-ordered now. Be sure to get yours before the transportation freezes over and you can't get it.

This wonderful game costs only $25,000. Yes, just twenty-five thousand dollars !

Perhaps that seems like a lot. But in fact money won't be worth anything after a while, and after all it's a tiny drop in the bucket compared to what the radical Left has been spending on writing fake research papers, and fighting a Global Warming that never materialized.

So order your game now, before the Post Office trucks freeze over.

Or alternatively, you can change your life political and social life, now that you have seen the light.

Vote Republican. Join he Republican party, and become an activist with them. Crush Democrat Party and JUSTICE DEMOCRAT voter fraud in our elections.

Get active in working to get America ready for the Ice Age which is on the doorstep.

You now can really do something to save the planet. It's not too late.

Even leading environmental organization founders and leaders are getting on the bandwagon against the Global Warming / Climate Change hysteria.

Pat Moore, Co-Founder of Greenpeace said in March 2019:

The Global Warming / Climate Change hysteria "is not only fake news. It's

fake science."

Carbon Dioxide is **"the main building block of all life"**. CO2 is good for the environment.

"There is nothing to be afraid of."

About the fake scientists mentioned in this book, he said they are going after: **"perpetual government grants," and insist "the science is settled and say people like myself should just shut up. On the other hand, they keep studying it forever as if there's something new to find out."**

He's right of course. And I might add, the fake Global Warming "scientists" have completely missed that the world is getting colder. A lot colder.

CHAPTER 31

A POLITICAL WAR AGAINST WE THE PEOPLE, HAS BEGUN. STARTED BY THE CLIMATE CHANGE POLITICAL LEFT. AND MAKE NO MISTAKE ABOUT IT, THEY HATE US.

IT'S ON. THE POLITICAL, JUDICIAL AND MEDIA ICE AGE WAR IS DECLARED

THE ICE AGE WILL BE HERE BETWEEN 2019 AND 2035 (around 2025), AND THE GLOBAL WARMING MOB IS HOLDING US BACK FROM PREPARING FOR IT

In mid-March 2019, the Global Warming lies mob showed up in full force. Well sort of. At least they tried to look like it.

Democrats, and the vipers they spawned, the JUSTICE DEMOCRATS / DEMOCRATIC SOCIALISTS rallied their mobs for a display of their BS.

This was all part of the "Children's Strike for Global Warming". In which little brainwashed and mind-numbed robots were wound up by their irresponsible parents, and sent out into the street.

Thus in this major offensive of lies, along with Gore, a few hundred brainwashed "children" protesting Global Warming slithered along the streets.

All hysterically pushing for action on a problem that does not exist. Global Warming.

Of course Climate Change has always existed. Billions of years. From the beginning of the earth until now. It always gets colder and warmer, colder and warmer.

But human activity and CO2 has nothing to do with it.

Climate Change which is Caused By Human Activity, is a fraud.

Thus, cute little children have been turned into cretins by adults. The brainwashing of children is completely disgusting and has to be met head on.

As you have seen, dear reader, the scientific inconvenient truth is that WE ARE HEADING INTO A NEW ICE AGE as the sun cools down.

In fact we are just emerging from the last Small Ice Age, which took place from 1300 to 1800. But now we are going down the frozen slippery slope of a NEW ICE AGE.

As usual, Al Gore told a series of lies in a horrifyingly false interview on March 15th 2019. It was difficult to watch, and in fact I had to fight against turning it off. But to see what the devil is doing, sometime we have to look at their messengers.

It's difficult to know where to start in cataloging the many lies he told in this one interview. But let's start with the mother of all their lies, that 97% of scientists support the idea that global warming is occurring and that it is caused by human activity.

As we have seen, it is much less than 1% of the "scientists" who support this Global Warming lie. And that those who do so, are doing it by false data, incorrect digital models, and fraudulent reports. Scammers.

Doing it basically for the money. They get paid a fortune to do their "research" and to publish their lies. Or in some cases for the political power motivation. But the money is probably foremost.

Paid by our own taxpayers' money through billions of dollars a year in government grants. And to make even more lies, the Global Warming fraud billionaires fund more studies which fit their agenda. The funders of mountains of written garbage include, but are not limited to, Soros, Steyer, Bezos, and Bloomberg.

Democrat Party government bureaucrats that can't be fired because of union agreements, along with the Climate billionaires, are radical leftists when it comes to Global Warming. They all fund false studies to further the Leftist political agenda of governmental tyranny.

Their message: because there is a fire coming because of Global Warming, of course we need an all powerful, dictatorial government to solve the problem.

They, in cooperation with the Democrat Party / JUSTICE DEMOCRATS, and the Democrat Media complex, are working together to give the government massive power over everything we do.

What we eat, how we travel, our health care, education for our children. Everything will be decided by the state. The individual will decide nothing. And descend into poverty and a new form of slavery. Unless stopped, the new slave masters will be the state and the Democrat Party.

This is why they all want to take guns away from people. It's again the Nazi thing. The Nazis were very tough on gun control, because they wanted their victims to be defenseless.

But getting back to the revolting TV interview. Gore was not content to let the lie rest at 97%. He said in this interview (and this is a paraphrase because I can not stand to watch that monster's video a second time to write it down exactly):

"The science is now settled! And it's 99% of the scientists who support Global Warming – Climate Change". Yikes, it's now 99%. Maybe next year 110%.

I'm leaving out of the paraphrase that during the entire interview, Gore was snickering, chortling in a sort of strange way: "Heh, heh, heh. Heh heh heh heh." Like a crazy person. Like he really is, in fact.

Perhaps he was repeatedly snickering because he is giddy. He thinks he is going to win his battle as a confidence trickster against truth. Or perhaps because he is mentally unstable. Or perhaps both.

Gore and all the rest of them, of course keep moving the goalposts when they are proven wrong.

First their hysteria was "Global Warming", and that didn't work out for them, because the weather has been freezing. Global Warming protesters were getting frostbite on their rear ends. They were told to dress for warm weather and it was below freezing.

Then the lie moved to from overheating to "Climate Change", something which has happened for billions of years on earth, and has nothing to do with human activity.

And that wasn't working so well, so they have moved to "Unusual Weather Events", caused by "Global Warming". Climate change, caused by us, human beings. Yikes.

Again the false statistics, the dramatic rants by everyone from slow moving, slow talking wooden dummy Chucky Schumer, to Frostbite Gore, to the street mob, and AOC... that these weather events are worse than ever.

First, this is not true. Stormy weather events have always been part of the planet. For billions of years. At time they wiped out entire areas of plants and animals. Vicious storms.

Anyway, in the rather awful interview, the so-called reporter, I don't recall her name, oh yes, Brooke, it seemed from her face and body language that she felt she had been made to dangle dead mice in front of a snake. Just saying what it looked like.

And of course she did some feeding.. by feeding to Gore leading questions which had the intent to attack the President of the United States. The team of Al and Brooke were truly revolting.

Because the President of the United States is correct on the climate issue, he is a target of the Democrat Media Complex, and the Global Warming mob.

The president is not just spouting off. He's getting the best scientific advice directly from academicians and 130 scientific organizations.

But back to the Brooke and Al horror movie.

The CNN interview truly like a horror movie. "Heh, heh, heh... heh, heh, heh" puked forth Frostbite Gore after almost every sentence.

CNN, or which ever channel it was, I do not remember or want to remember, could not hide the evil behind the lies. And as mentioned, even the woman doing the interview looked as if a snake was crawling around her toes. She really looked awful.

It was similar to watching a Halloween program. But Halloween in reverse. In this case many many real monsters, were dressing up like humans.

Many many because in addition to the horror film interview, the so called "children's strike" for climate action was also dutifully covered by the Democrat Media Complex.

Yet out of the total number of children who could have participated, only a tiny fraction did. Out of every 1 million children, 999,999 kids did NOT show up. Out of possible tens and tens of millions. A few hundred in each place showed up. Together maybe a few thousand at most.

But of course the Democrat Media Complex and the Global Warming mob which is at war with America, presented it as *all children*.

But isn't even just brainwashing one child a crime? Turning them into mind-numbed, arrogant little robots.

It's probably illegal to drive children insane. Now is the time to have the criminals of the Global Warming fraud called to account. This means the Democrat Party, the JUSTICE DEMOCRAT / Democratic Socialist group, the Democrat Media Complex, the billionaires and so on.

Called to account by our republican constitutional action groups through actions in the political, judicial, and media areas.

What the tyrannical Left has done, in politics and in environmental movements, is to create children who are intolerant bigots. Cute little children have been turned into climate Stalins and Hitlers.

Children of the so-called "Sunrise" group, are examples. Like little fountains of vomit, they spout lies they have been spoon fed by the Global Warming hysteria mob. They've been made into arrogant kids. Dumb and filled with lies. In other words, into typical Democrats.
In late February 2019, Senator Diane Feinstein, was ambushed by a bunch of kids and a few adults from the Sunrise organization.

The children were arrogant and referred to the senator in the conversation as "Feinstein", and not "Senator Feinstein".

Now I do not like the politics of Diane Feinstein. But those children and their puppet master adults were simply little pieces of rude and arrogant puke.

In addition to to parents producing daddy and mommy mind-numbed carbon copies of themselves, the schools and universities are doing their part to increase the number of little mind-numbed robots. Brainwashing is a crime.

Kids, wake up! Parents save your children! And if you are a parent who has been contributing to this, put yourselves into rehab and turn your children over to loving, caring, rational constitutional conservative relatives who love them.

SCIENTISTS DO AGREE, BUT IT'S NOT AGREEMENT TO WHAT THE GLOBAL WARMING MOB THINKS

Leading scientists in relevant fields (solar science, astrophysics, physics, chemistry, cosmology, geology climatology, mathematics, engineering and on and on) do agree on this:

AN ICE AGE IS AT THE DOOR STEP

I first wrote about this in a paper debunking Global Warming in 2009. At that time, my estimate, based on the activity of the sun and projections about the coming solar minimum, was about 15 years from then. In other words about 2024.

Now there is much better data available than I had in 2009, to estimate then the Ice Age will begin, and scientists from NASA are contributing as well.

The agreement is this: the ice age hits between 2019 and 2035. Very likely the due date is 2025.

I'm not patting myself on the back, but somehow I got at least slightly close. At that time it was mostly Japanese and Russian scientists who predicted the coming ice age, with several also in the US.

THE TRUTH ABOUT GLOBAL WARMING...

Southern California's tropical desert covered in snow December 2008

Dr. Joel Glass

September, 2009

GLOBAL WARMING IS THE GREATEST SCIENTIFIC FRAUD IN HISTORY

The world is entering a period of dangerous global cooling, a new mini-ice age.

Just for fun here are, a few pages from my original paper, which include blowing up Gore's totally fake "hockey stick" graph, which fooled so many people in 2006,

> **DEFINITION OF "GLOBAL WARMING"**
> **THE FRAUDLENT SCIENCE IN BRIEF:**
>
> - the belief that the earth is dangerously warming
> - that human activity is the cause of the warming
> - and that radical steps that would cause huge reductions in standards of living around the world are required to save the planet from extinction
>
> Each one of these three suppositions are junk science completely wrong.
>
> Former Alaska Governor Sarah Pallin, who during her vice presidential campaign in the United States, said that global climate change was not man-made, was absolutely right.
>
> **THE "HOCKEY STICK ANALYSIS", A BASIS OF THE GLOBAL WARMING FRAUD DRAMA, a basis of global warming for Al Gore, the United Nations, and other alarmists**
>
> Geological scientists Michael E. Mann, Raymond S. Bradley and Malcolm K. Hughes have created (1998 – 2001 for use in a UN report) a graph of what they claim shows temperature changes of the last 1000 years, with massive global warming during the last

and is still fooling them today. It's a complete fake:

In looking at my original paper today while writing this chapter, I found a few nice quotations which I had forgotten about but which are important for us.
Some of these were a confrontation to the UN fake climate change agenda, and were presented to the International Climate Conference in Poland in 2008.

"UN Blowback:
More Than 650 International Scientists Dissent Over Man-Made Global Warming Claims POZNAN, Poland

"- The UN global warming conference currently underway in Poland is about to face a serious challenge from over 650 dissenting scientists from around the globe who are criticizing the climate claims made by the UN IPCC and former Vice President Al Gore.

Set for release this week, a newly updated U.S. Senate Minority Report features the dissenting voices of over 650 international scientists, many current and former UN IPCC scientists, who have now turned against the UN.

The report has added about 250 scientists (and growing) in 2008 to the over 400 scientists who spoke out in 2007.

The over 650 dissenting scientists are more than 12 times the number of UN scientists (52) who authored the media hyped IPCC 2007 Summary for Policymakers."

http://www.m4gw.com:2005/m4gw/

Here are some of the statements by these world leading scientists who ridicule Global Warming as the greatest scientific fraud in history:

"I am a skeptic...Global warming has become a new religion."

- Nobel Prize Winner in Physics, Ivar Giaever.

"Since I am no longer affiliated with any organization nor receiving any funding, I can speak quite frankly....As a scientist I remain skeptical."

- Atmospheric Scientist Dr. Joanne Simpson, the first woman in the world to receive a PhD in meteorology and formerly of NASA who has authored more than 190 studies and has been called "among the most preeminent scientists of the last 100 years."

"Warming fears are the "worst scientific scandal in the history...When people come to know what the truth is, they will feel deceived by science and scientists."

- UN IPCC Japanese Scientist Dr. Kiminori Itoh, an award-winning PhD environmental physical chemist.

"The [global warming] scaremongering has its justification in the fact that it is something that generates funds."

- Award-winning Paleontologist Dr. Eduardo Tonni, of the Committee for Scientific Research in Buenos Aires and head of the Paleontology Department at the

University of La Plata.

http://www.epw.senate.gov/public/index.cfm?
FuseAction=Minority.Blogs&ContentRecord_id=2158072e802a-23ad-45f0-274616db87e6

"Gore prompted me to start delving into the science again and I quickly found myself solidly in the skeptic camp...Climate models can at best be useful for explaining climate changes after the fact."

- Meteorologist Hajo Smit of Holland, who reversed his belief in man-made warming to become a skeptic, is a former member of the Dutch UN IPCC committee.

"Many [scientists] are now searching for a way to back out quietly (from promoting warming fears), without having their professional careers ruined."

- Atmospheric physicist James A. Peden, formerly of the Space Research and Coordination Center in Pittsburgh.

"Creating an ideology pegged to carbon dioxide is a dangerous nonsense... The present alarm on climate change is an instrument of social control, a pretext for major businesses and political battle. It became an ideology, which is concerning."

- Environmental Scientist Professor Delgado Domingos of Portugal, the founder of the Numerical Weather Forecast group, has more than 150 published articles.

"The IPCC (the UN climate study group) **has actually become a closed circuit; it doesn't listen to others. It doesn't have open minds... I am really amazed that the Nobel Peace Prize has been given on scientifically incorrect conclusions by people who are not geologists,"**

- Geologist Dr. Arun D. Ahluwalia at Punjab University and a board member of the UN-supported International Year of the Planet.

"The models and forecasts of the UN IPCC "are incorrect because they only are based on mathematical models and presented results at scenarios that do not include, for example, solar activity."

- Victor Manuel Velasco Herrera, a researcher at the Institute of Geophysics of the

"Even doubling or tripling the amount of carbon dioxide will virtually have little impact, as water vapour and water condensed on particles as clouds dominate the worldwide scene and always will."

– . Geoffrey G. Duffy, a professor in the Department of Chemical and Materials Engineering of the University of Auckland, NZ.

"After reading [UN IPCC chairman] Pachauri's asinine comment [comparing skeptics to] Flat Earthers, it's hard to remain quiet."
- Climate statistician Dr. William M. Briggs, who specializes in the statistics of forecast evaluation, serves on the American Meteorological Society's Probability and Statistics Committee and is an Associate Editor of Monthly Weather Review.

"For how many years must the planet cool before we begin to understand that the planet is not warming? For how many years must cooling go on?"

- Geologist Dr. David Gee the chairman of the science committee of the 2008 International Geological Congress who has authored 130 plus peer reviewed papers, and is currently at Uppsala University in Sweden.

"It is a blatant lie put forth in the media that makes it seem there is only a fringe of scientists who don't buy into anthropogenic global warming."

- U.S Government Atmospheric Scientist Stanley B. Goldenberg of the Hurricane Research Division of NOAA.

All these have been hidden from you by AOC and her little friends at the Justice Democrats / Democrat Socialist coven.

Hidden by Al Gore. By the Sunrise group. By the Leftist billionaires. By the Democrat Party and by the Democrat Media Complex.

This book doesn't have the possibility to go into the role of the Democrat Media Complex in the Global Warming fraud.

But let's just say, the Democrat Media Complex is part and parcel of the Nazi propaganda methodology. Willing servants of the tyrannical left. Here is what their idol and model, Joseph Goebbels had to say:

"A lie told once remains a lie but a lie told a thousand times becomes the truth"

and

"Think of the press as a great keyboard on which the government can play."

Yes indeed, the Democrat Media Complex is one big blatant lie. And they have hidden these statements from leading scientists behind their shadows of darkness at CNN, MSNBC, ABC, CBS, NBC, The Washington Post, The New York Times, NPR, The Associated Press, Bloomberg Media, Politico, Buzzfeed and so on. Following the Nazi propaganda playbook, whether they know it or not.

These statements above by leading scientists are the truth. The Democrat Media Complex never reports this. And reports only lies as if they were true.

The Democrat Media Complex is not really media or news. But a propaganda machine to push forward the fake Global Warming agenda. These Democrat Media Complex companies have a heavy guilt for spreading the climate and other political lies, particularly now.

With ignorant fools such as F. Chuck Todd (who managed managed to pull together the mental resources to sign a legal document changing his name to "Chuck Todd", and dropping the "F"), they are stumbling over the ice age cliff.

For the damage they have done to leading American scientists. For the damage they have done to the President and to the nation by pushing forward the fake Global Warming / Climate Change / Catastrophic Weather Events agenda, they need to be held responsible in the courts of justice.

Get the attorneys ready. More and more cases related to lies, such as the existing $250 million defamation case against the Washington Post, and the $270 million defamation case against CNN. These point the way. The judicial way.

Just a word about a sad recent political development, which we hope to reverse... After fighting the good fight since the mid-1980s against the Global Warming lie, many if not most Republicans in congress are waving the white flag of surrender.

All except the Freedom Caucus, in the House and constitutional conservatives in the Senate. Many Republican law makers are giving up the fight.

Not because they have been convinced Global Warming / Climate Change is occurring and is caused by human activity, but because they are tired. They can't seem to make headway in the struggle for truth. So they are going into shut-down mode.

This is dangerous for the country, because the Global Warming / Climate Change mob is pushing the United States of America into national suicide.

We are being forced to surrender liberty and freedom to the government.

>Our lives to be controlled by ignorant fools such as AOC, Swalwell, Omar, and the other power-hungry totalitarian Democrats controlled in the end by their billionaire backers.

In addition, we will be spending the pitifully small remaining amount of America's wealth to solve a problem that does not exist: CO_2 causing Global Warming and severe weather events. Events that have always been with us since the beginning of time.
We might as well spend tens of billions of dollars a year to find the tooth fairy.

And we are losing the time and resources to get ready for the real disaster that looms right in front of our faces. THE COMING ICE AGE

America, and the world in fact, should be getting our water delivery systems, our roads, our vehicles, our power and energy industry ready for frozen conditions. But we are not doing it.

Thus, when the ICE AGE hits rather soon, as we mentioned, about half the US population will die in a few years. Their blood is on the hands of the Global Warming / Climate Change leaders, the Democrat Party, the so-called JUSTICE DEMOCRATS / DEMOCRATIC SOCIALISTS, and the corrupt insane Democrat Media Complex.

IT'S TIME NOW TO FIGHT BACK

IT'S NO LONGER ENOUGH TO SIMPLY PRESENT REAL FACTS, REAL SCIENTISTS. WHICH PROVIDE THE INCONVENIENT TRUTH THAT GLOBAL WARMING / CLIMATE CHANGE IS A FRAUD.

IT'S TIME TO NAME AND SHAME THE ICE AGE DENIERS.

TO ISOLATE AND DESTROY THEM POLITICALLY, IN THE JUDICIAL SYSTEM, AND BY THE MEDIA.

THIS IS A WAR FOR SURVIVAL, OF THE PEOPLE, BY THE PEOPLE AND FOR THE PEOPLE.

The new activist movement against the ICE AGE DENIERS has begun. Join your local Republican Party and work from within.

Here is some chilling spears of truth to hurl at the ICE AGE DENIERS

A "Mini Ice Age" Is Coming Soon Says Math Professor's Solar Cycle Model That's 97% Accurate

The most recent research to examine this topic comes from the National Astronomy Meeting in Wales, where Valentina Zharkova, a mathematics professor from Northumbria University (UK), presented a model that can predict what solar cycles will look like far more accurately than was previously possible.

She states that the model can predict their influence with an accuracy of 97 percent, and says it is showing that Earth is heading for a "mini ice age" in approximately fifteen years.

According to the Royal Astronomical Society (RAS):

A new model of the Sun's solar cycle is producing unprecedentedly accurate predictions of irregularities within the Sun's 11-year heartbeat.

The model draws on dynamo effects in two layers of the Sun, one close to the surface and one deep within its convection zone.

Predictions from the model suggest that solar activity will fall by 60 per cent during the 2035s to conditions last seen during the 'mini ice age' that began in 1645.

Zharkova and her team came up with the model using a method called "principal component analysis" of the magnetic field observations, from the Wilcox Solar Observatory in California.

Looking forward to the next few solar cycles, her model predicts that from 2035 to 2040 there will be cause for a significant reduction in solar activity, which again, will lead to a mini ice age. According to Zharkova:

"In cycle 26, the two waves exactly mirror each other – peaking at the same time but in opposite hemispheres of the Sun. Their interaction will be disruptive, or they will nearly cancel each other.

We predict that this will lead to the properties of a "Maunder minimum."

Effectively, when the waves are approximately in phase, they can show strong interaction, or resonance, and we have strong solar activity. When they are out of phase, we have solar minimums.

When there is full phase separation, we have the conditions last seen during the Maunder minimum, 370 years ago.

https://www.sciencedaily.com/releases/2015/07/150709092955.htm

We believe this is coming a bit earlier than 2035, we believe according to and along with NASA scientists that could start soon.

But whichever is correct 2020, 2035 or in between, we are about to start the decent into frozen hell.

Take a look at the future:

Isn't it time to take the fight where it should be: right into the face of the ICE AGE DENIERS

ICE AGE DENIERS

It's projected to come in 2020 or around 2035, take your choice on the new ice age arrival.

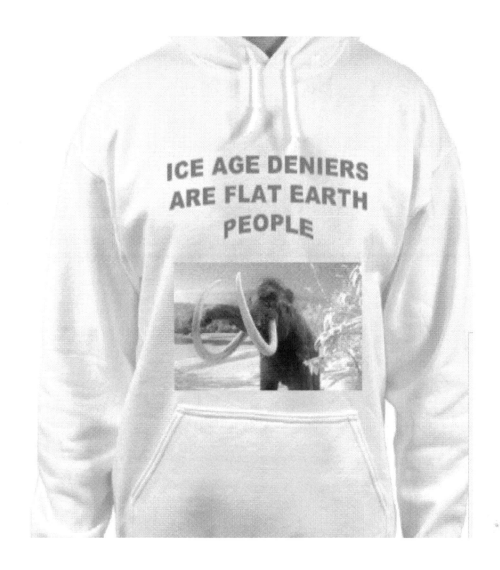

That gives us just a little time to get our country, our families ready for the NEW ICE AGE.

And that means sweeping the Global Warming / Climate fraud leaders, and the Democrat Media Complex into the dustbin of history.

And helping the victims they have brainwashed to regain their senses, their freedom, and their sanity.

LET'S ... GET TO THE ROOTS

OF WHAT HAS CAUSED THE DISPICABLE GROUP BASED ON LIES, THE GLOBAL WARMING / CLIMATE CHANGE MOVEMENT

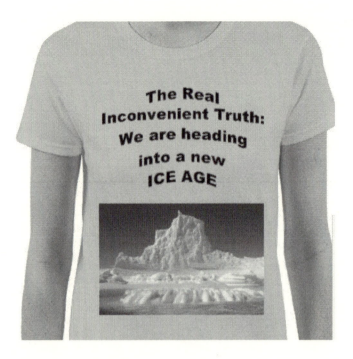

It's a poisonous weed. For the health of liberty, life and the pursuit of happiness, we need to get rid of it.

Not only for the country but even for the health of the people who have been brainwashed by it.

They can be saved too.

The so-called Global Warming movement is not about climate at all, but about a lust for tyrannical power.

Democrat Party, JUSTICE DEMOCRAT, DEMOCRAT SOCIALIST Global Warming programs are less about climate and all about grabbing the rights of American Citizens.

They really could care less about the climate. It's all about power. 12 years left, means that they want We the People to surrender liberty, freedom, and even life, so they can grab absolute power.

The Global Warming mob objective is the destruction of:

Freedom of Speech
Freedom of Religion

**Freedom of Assembly
Freedom of Political Choice
Freedom of the Press**

**The Right to Bear Arms
Life
Liberty
Pursuit of Happiness**

And that's just for starters.

With Freedom of the Press gone, the only thing you could watch on TV is CNN. Can you imagine 24 hours a day of Jim Acosta?

These rights enumerated in the constitution, are all the targets of the Democrat / JUSTICE DEMOCRAT / DEMOCRAT SOCIALIST tyrannical Global Warming program.

To put it bluntly, it's their way to establish a Stalinist-type totalitarian state in America. With them at the wheel. Yikes.

To shred the constitution and its Bill of Rights.

And put them in control of every aspect of your life. How you travel, what you eat. Where you live. Everything. Freedom gone.

The Democrat totalitarian program to destroy the rights of every American except for their "elite", can be seen in an excellent Twitchy article, and we need not go into too much detail here. It's clear because the Democrats admit it.

https://twitchy.com/brettt-3136/2019/03/16/alyssa-milanos-tweet-makes-it-clear-the-fight-over-climate-change-has-nothing-to-do-with-climate-change/

So what's really at the root of the Global Warming / Climate Change lies?

It's above all, at totalitarian movement. With the objective to destroy human rights.

Love of money, by the leaders and lying scientists

Lust for power over others. A drive for absolute power to destroy individual liberty. In short, it's a totalitarian movement. No different from that of Nazi Germany or Soviet Russia.

It's satanic. A false religion. It's a new Church of Satan.

> One of the best books to refer to about Global Warming is the bible. There it says clearly:
>
> SATAN IS THE FATHER OF LIES

... So if you want to know where the Global Warming lies come from, look no further than that.

The father of the Global Warming / Climate Change lies is not actually Al Gore, but is who is in Al Gore and the Global Warming leaders who followed him. Puppets of Satan, the false accuser.

The Democrat Media Complex, is absolutely Satanic.

They lie without cease. Lie, after lie, after lie.

These are not normal people. They are at war with America. They are demonic.

If people don't like the term "demonic", no problem. We can compared the Democrat Media Complex global warming campaign to something else.

How about rabid dogs. Running around, often in circles, barking and snarling. And telling endless lies.

And they can bite with very serious consequences.

The virulent, dishonest attacks by the Democrat Media Complex on President Trump. And the Democrat Media Complex lies on Global Warming, are not the first time this media has been at war with the American people.

The New York Times and the Washington Post, were essentially Holocaust deniers during the Second World War. They basically hid information about the mass murder of Jews from the American public, and also from the US government.

They had the information about the genocide, but either didn't report it at all. Or they buried it in small articles once in a while, on back pages.

The Left wing rabid dog media, was and is... despicable.

This is of course heavy stuff. But We the People, the vast majority of not only the US, but the world, need to understand what we are up against.

This is a movement as satanic as Nazism, or Soviet totalitarianism. And it is modeled on both.

THE BIG LIE

Absolute Power of A Small "Elite" over We the People.

But it isn't going to happen.

ORGANIZE

POLITICALLY, JUDICIALLY, AND IN THE OUR OWN MEDIA EFFORT

In addition to possible law suits against the Democrat Media Complex for damages and spewing lies without cease, why not in the courts hold responsible the scientists who continue to lie for money or political power?

The conservative blogger, Needs of the Many wrote:

"It is also time that we launch massive investigation into the scientific community to find out which scientist continued to lie and manipulate data about global warming in order to get funding.

You should already know the huge amounts of money involved with supporting global warming arguments through funding (over $80 billion).

While those scientists who have knowingly lied about global warming in order to get money are being punished in the wheels of the justice system ...

we should reward the huge numbers and majority of published scientist that were and are honest by saying man is not responsible, or that there was not enough evidence to say either way.

http://needsofthemany.wordpress.com/2008/03/04/global-warming-is-officially-over-suck-

itgreenies-2/

Yes, and why not take legal action against the false religion guru himself, and many of his followers that have supported him with fake scientific studies, Al Gore. Fraud is a crime, and if it is proven in the courts, let the chips fall where they may.

Even forgetting his 20 years of being a constant liar, the single statement on his March 15, 2019 TV interview, that "99% of scientists" support the Global Warming / Climate Change being caused by humans scenario, may be a good start to pursue possible judicial action. It's a con man's statement.

The hundreds of millions of dollars in assets he has gained through a career of lies could be removed with just cause by judicial decisions for damages and fraud.

Similarly, the organizations, such as the Democrat Party, the Justice Democrats / Democrat Socialists, and the billionaires who fund them, are all potential juridical targets for law suits.

Much information in this book could be possibly be legal ammunition for the attorneys acting on behalf of truth, reason, and real justice.

And dear readers, the fight forward is not only in the United States. As mentioned in this book, world-wide people are waking up.

For example, in Japan:

A scientific commission of the prestigious Japan Society of Energy and Resources, called the Global Warming hypothesis "unprovable"

http://newsbusters.org/blogs/noelsheppard/2009/02/25/japanese-commission-challenges-un-global-warming-not-man-made

The leading scientists on the Japanese commission contend that recent climate change is driven by natural cycles, not human industrial activity, as political activists argue.

Kanya Kusano is Program Director and Group Leader for the Earth Simulator at the Japan Agency for Marine-Earth Science & Technology (JAMSTEC).

He focuses on the immaturity of simulation work cited in support of the theory of anthropogenic (caused by humans) climate change.

Using undiplomatic language, Kusano compares them to ancient astrology.

After listing many faults, and the IPCC's own conclusion that natural causes of climate are poorly understood, Kusano concludes:

"[The IPCC's] conclusion that from now on atmospheric temperatures are likely to show a continuous, monotonic increase, should be perceived as an unprovable hypothesis,"

Shunichi Akasofu, head of the International Arctic Research Center in Alaska, has expressed criticism of the theory before. Akasofu uses historical data to challenge the claim that very recent temperatures represent an anomaly:

"We should be cautious, IPCC's theory that atmospheric temperature has risen since 2000 in correspondence with CO2 is nothing but a hypothesis."

Akasofu calls the post-2000 warming trend hypothetical. His harshest words are reserved for advocates who give conjecture the authority of fact.

"Before anyone noticed, this hypothesis has been substituted for truth...

(Now) The opinion that great disaster will really happen (because of Global Warming) must be destroyed."

http://www.theregister.co.uk/2009/02/25/jstor_climate_report_translation/

And here is some more for the wonderful lawyers that are going to bankrupt the lying scientists and media who sending the US into national suicide:

THE COMPUTER CLIMATE MODELS, CREATED EITHER OUT OF IGNORANCE OR BY PURPOSELY DISTORTING VARIABLES AND DATA...

THE SAME MODELS USED BY THE UNITED NATIONS, ARE FUNDAMENTALLY FLAWED AND INACCURATE

Dr. John S. Theon, former Chief Atmospheric Scientist at NASA, who was in charge of all climate related research, and who was the supervisor of Global Warming Fraud perpetrator Hansen, has said that Hansen acted without authorization and against the actual data which was at NASA.

http://epw.senate.gov/public/index.cfm?FuseAction=Minority.Blogs&ContentRecord_id=1a5e6e32-802a-23ad-40ed-ecd53cd3d320

Theon is a global warming skeptic, and believes that the climate models in use by the United Nations and other global warming advocates are deeply flawed and inaccurate: "climate models are useless.

My own belief, concerning anthropogenic climate change is that the models do not realistically simulate the climate system because there are many very important sub-grid scale processes that the models either replicate poorly or completely omit,"

"Furthermore, some scientists have manipulated the observed data to justify their model results.

In doing so, they neither explain what they have modified in the observations, nor explain how they did it.

They have resisted making their work transparent so that it can be replicated independently by other scientists. This is clearly contrary to how science should be done.

Thus there is no rational justification for using climate model forecasts to determine public policy,"

"As Chief of several of NASA Headquarters' programs (1982-94), an SES position, I was responsible for all weather and climate research in the entire agency, including the research work by James Hansen, Roy Spencer, Joanne Simpson, and several hundred other scientists at NASA field centers, in academia, and in the private sector who worked on climate research. This required a thorough understanding of the state of the science."

http://epw.senate.gov/public/index.cfm?FuseAction=Minority.Blogs&ContentRecord_id=1a5e6e32-802a-23ad-40edecd53cd3d320

The Democrat Party political left, and their media and fraudulent Global Warming hysterics hate us. And they are ICE AGE DENIERS, leading the country into national suicide.

ICE AGE DENIERS

There is substantial evidence that the Global Warming / Climate Change movement is a fraud. One causing serious damage.

Fraud is a crime in the United States. Global Warming / Climate Change Caused by Human Activity is clearly a fraud. One created for financial and for political power purposes.

FRAUD... THEY DID IT... NOW THEY PAY

Fraud is legally defined as deliberately deceiving someone else with the intent of causing damage. This damage need not be physical damage, in fact, it is often financial.

And is not financial gain the underlying reason for all the false research and science? The corruption of data. The published papers which are lies and scientific crap?

The precise legal definition of fraud varies by jurisdiction and by the specific fraud offense.

Fraud consists of some deceitful practice or willful device, resorted to with intent to deprive another of his right, or in some manner to do him an injury.

Here we go, because a main purpose of the Global Warming / Climate Change fraud is to deprive American citizens of their rights and freedom under the constitution.

Fraud includes all acts, omissions, and concealments which involve a breach of legal or equitable duty, trust, or confidence justly reposed, and are injurious to another, or by which an undue and unconscientious advantage is taken of another.

Aren't the Global Warming actions of the Democrat Party, JUSTICE DEMOCRATS / DEMOCRAT SOCIALISTS in congress and state governments a breach of legal or equitable duty and trust?

Fraud is deliberate deception to secure unfair or unlawful gain, or to deprive a victim of a legal right.

And that is what the Global Warming / Climate Change mob wants to do.

In government, in the media, in political parties. They want to deprive the American people of natural rights enumerated in the Constitution of the United States.

Getting back to the basics of the fraud behind the power grab.

Climate Change and Global Warming or Global Cooling are caused by the sun. The amount of solar energy the sun is producing. Anything else said about it is fraud.

Dr. Jay Lehr, a leading American scientist, was asked by Lou Dobbs what he considered the dominant influence on Earth's climate.

"Well, clearly, Lou, it is the sun," Lehr answered, adding that "if we go back in really recorded human history; in the 13th century, we were probably seven degrees Fahrenheit warmer than we are now."

Lehr considers global cooling to be the real threat, part of a natural pattern as we continue coming out of a period known as the Little Ice Age.

"If we go back to the Revolutionary War, 300 years ago," he said, "it was very, very cold. We've been warming out of that cold spell from the Revolutionary War period. And now we're back into a cooling cycle."

The Associated Press, part of the Satanic media, for example works to distort the truth on climate day and night. For example the AP claims that the 10 warmest years on record have occurred since Bill Clinton's second inaugural.

But after it was discovered that NASA's James Hansen, Gore's chief scientific ally, had been fudging the numbers, NASA was forced to correct its data.

The 10 warmest years turn out to be, in descending order: 1934, 1998, 1921, 2006, 1931, 1999, 1953, 1990, 1938 and 1939.

http://www.ibdeditorials.com/IBDArticles.aspx?id=314582265558716

Where we are headed is not a hot dry desert, it is this. A colder and colder world. Take a look at AOC's New York City in the near future:

The Global Warming / Climate Change movement is a fraud. And in keeping the United States from preparing for the coming Ice Age, they are causing incalculable damage for financial and political power gain...

THE DEMOCRAT PARTY /JUSTICE DEMOCRATS / DEMOCRATIC SOCIALISTS... PLUS THE DEMOCRAT MEDIA...

ARE USING NAZI PROPAGENDA METHODOLOGY AS PART OF THE FRAUD.

Joseph Goebbels, their apparent master, taught:

"Accuse the other side of that which you yourselves are guilty".

Gore and his present day sycophants, particularly Democrat Party / JUSTICE PARTY / DEMOCRATIC SOCIALISTS members of congress, **are constantly attacking what they call "Climate Deniers".**

THE CLIMATE DENIERS FRAUD

This is an insane term. Meant to cover up for it getting colder and colder. So the Global Warming movement attempts to silence the truth by smearing them.

No one in this world denies climate. Climate of course exists. The term is a smear. It shows how the thought police of the tyrannical Left tries to control the language.

But none the less the Stalinist-type, Nazi-type propaganda machine of the Global Warming movement and the Democrat Party / Justice Socialists / Democratic Socialists, is constantly accusing those who disagree with their false data and intentional lies, as doing something wrong.

If there are criminals and totalitarians involved in climate issues, it is not the scientists and the Republicans who are speaking the truth.

This Democrat Party, Justice Democrats / Democrat Socialist and the few remaining fraudulent academics who peddle the Global warming drug, have been using tactics straight out of a German totalitarian playbook. They call others "Climate Deniers", or "Global Warming Deniers", or "Climate Change Deniers". While they are actually denying the truth right under their noses that the Sun controls climate on earth. And that we are heading into a cold period.

But they stick with the Nazi propaganda playbook.

It is astonishing, that the JUSTICE SOCIALIST / DEMOCRAT propaganda is a carbon copy of the program utilized and theorized about by Nazi Minister of Propaganda, Joseph Goebbels. If Goebbels taught it, the JUSTICE DEMOCRATS / DEMOCRAT PARTY are doing it. Their Nazi roots of propaganda and political power grab run

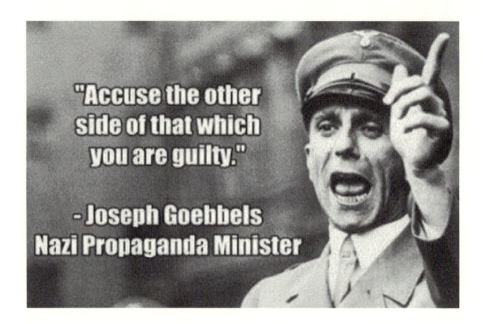

deep:

When it is they themselves who are denying the truth. They are, if we may say with all due respect:

ICE AGE DENIERS

It is very possible, that if they could, the Global Warming / Climate Change mob would round up those who disagree with them and put them in concentration camps.

Democrat president FDR did it to the Japanese, and the Democrat Party has compared Global Warming / Climate Change to World War II. One can imagine what they might do if the Global Warming hysterics felt unfettered.

But the President, the patriotism of law enforcement, the patriotism of the military, and the Second Amendmen in the United States is standing in the Way of the Democrat Party and the little viper parties they have spawned.

Let's be honest about it, the Second Amendment is a major factor in the protection of rights and liberty from the totalitarian Left.

Make no mistake about it, if they could, the Global Warming maniacs would imprison anyone who disagrees with their lies.
We the People are not suggesting that. We are suggesting that if the crime of fraud has occurred, and is proven in a court of law, then the Global Warming movement will face the consequences.

They do not know apparently, that we are just getting out of the last small-ice age, and are about to slide into the next one.

This next ice age may be small or huge. But even small it causes massive death, collapse of societies, and rampant lawlessness and violence. As well as the death of animals and plants. All this if we do not prepare for it.

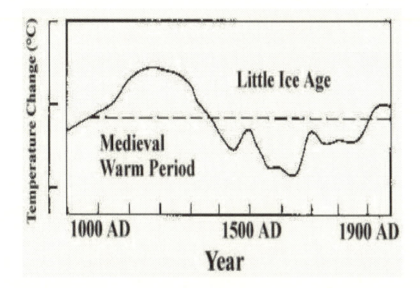

As we see from this chart, which we can introduce to our friends and relatives who are part of the Global Warming mob, as "like a piece of wet cooked spaghetti on a piece of paper... just follow the wet spaghetti on its path down into the freezer."

The world is just emerging from a recent Mini-Ice Age. That occurred between 1300 and 1800 or 1850 (as mentioned previously in this book). But we are now making a downward dip into a new mini-ice age.

And those who push the totally discredited Global Warming / Climate Change By Human Behavior fraud, are actually pushing the country into national suicide.

A double suicide in fact.

First their efforts to take and destroy the natural rights of citizens enumerated by the Constitution.

And Secondly by keeping the US from preparing for an ice age, they are setting the stage for catastrophe .

THE DEMOCRAT PARTY / JUSTICE DEMOCRAT GLOBAL WARMING MOB IS LIKE THE WALKING DEAD

The Global Warming / Climate Change movement, seems to never stop. No matter how discredited their falsified data, fake models, failed hysterical projections of disasters, and political lies become.

They are like the walking dead. The just stumble forward in a quest for power over We the People, who life in a world of Life, Liberty and Happiness, under heavenly-given rights.

The difference between the Global Warming leaders with their mob and the walking dead... is that Al Gore and his mob they have not come to get us like zombies to eat our brains, but as a colleague of mine said, they simply come and hope we don't have any.

But because we do, if we fight back with focus and power, then as Ronald Reagan said: We win, they lose.

Take a look at our victory T-shirt:

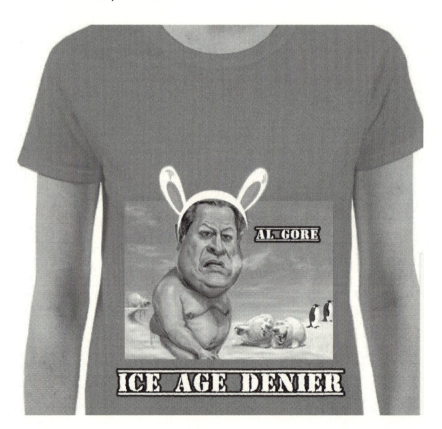

CHAPTER 32

IS THE GLOBAL WARMING MOVEMENT REALLY SATANIC ?

LET'S USE THE RED CHECK LIST AND FIND OUT

To determine if the Global Warming / Climate Change From Human Activity movements are satanic, we have to clearly define, and in detail, what satanic means.

The word "Satan" is not just a name without meaning. It has a meaning. And that meaning in detail is: THE FALSE ACCUSER

In the book of Job, that is what Satan does. He falsely accuses Job of all kinds of terrible things.

Then, the bible gives further understand of what satan is. It says:

- Satan comes to kill, steal, and destroy

So we have a false accuser, that comes for a purpose. And that purpose is to kill, steal and destroy.

Now we can determine whether or not the Global Warming / Climate Change movement is satanic in nature.

#1 A FALSE ACCUSER

The movement has accused anyone who disagrees with it, and from the very start, of being a sort of criminal. A science denier. A climate denier !

It has sought to smear and ridicule anyone who disagreed with their disproven theories.

Their accusation rhetoric has only ratcheted up over time. And now anyone disagreeing with the false data and false science of Global Warming / Climate Change is called a danger.

No doubt, were it not for the constitution, the movement would attempt to lock people up who disagree with them, or do worse.

So on point number one we can place a RED CHECK.

Yes the movement is a FALSE ACCUSER

2 HAS IT HAS COME TO KILL?

Their first victim is the truth.
They have tried to kill it from the start.

The Global Warming hysteria is a military-like effort to kill the truth, and has several components. One is the political parties themselves of the Global Warming movement.

These include, but are not limited to, the Democrat Party, and THE JUSTICE DEMOCRATS / DEMOCRATIC SOCIALISTS. And the Democrat Media Complex.

They lie on an hourly basis, and try to stifle any discussion or debate.

The Democrat Media Complex mouthpiece includes:

CNN, MSNBC, ABC, CBS, NBC

The political Left of the so-called White House Press Corps (which is about 60 out of 63 surveyed)

National Public Radio, National Public Broadcasting, The Associated Press, Bloomberg Media

The New York Times, The Washington Post

Internet sites such as BuzzFeed, Politico, Huff Post and so on

And the corporations which own them. Not only are these media corporations spewing the Global Warming lies endlessly, they attack any other media which dares to speak the truth. Fox News is their special target. Anyone who dares to challenge their distortions, lies, and omissions must be destroyed.

F. Chuck Todd stated on his NBC Meet the Press program:

We are "not going to give time to climate deniers," and went on to inaccurately characterize the nature of the climate debate.

"Just as important as what we are going to do is what we're not going to do," he said. "We're not going to debate climate change, the existence of it. The earth is getting hotter and human activity is a major cause. Period."

"We're not going to give time to climate deniers," Mr. Todd added. "The science is settled even if political opinion is not."

In addition to Chuck's statement being idiotic (keep it up Chuck), it's satanic.

The Spanish Inquisition was Satanic, The Cult of Stalin in Soviet Russia was Satanic, the Nazi movement was Satanic. The Democrat Party today is Satanic. For trying to silence the opposition. To falsely accuse the opposition of what they themselves are guilty of.

In addition to that., by their insane radical environmentalism, many people, animals and trees have been killed.

For example, the forest fires in California during the summer of 2018, were caused by lack of forest management, which left millions of dead trees and shrubs in the forest. This acted as a tinder box for the fire. That's the local level effect. On the national, it's just as bad.

By pushing the national policy of the United States, and in fact the world, towards solving a problem that does not exist, the Global Warming hysterics are killing coming generations. There is no effect of CO_2 on climate and no effect of human activities on climate change.

What is coming is a colder time period. A much colder time period. And because we will not be ready for it. We are not preparing the water, transportation, food and energy infrastructure to deal with the coming cold... this Global warming mob will have the deaths of millions on their hands.

It's apparently not enough for the Democrat Party, JUSTICE DEMOCRATS / DEMOCRAT SOCIALISTS for a few people to be killed in a massive forest fire. **They want open borders to really ramp up the action against American citizens.**

About 1 million to 1.5 million illegal aliens enter the country every year. We don't know who they actually are, or where they are.

Out of that million, tens of thousands are criminals. People with criminal records, people who have come with intent to do harm.

It is estimated that tens of thousands of US citizens are injured or killed every year by illegal aliens.

Specific estimates are that over 9,000 Americans are killed every year by

illegal aliens in the act of a felony.

Global Warming and Open Borders go together. They are part of the "Social Justice" are handmaidens of the lunatic Democrat Party and the JUSTICE Democrat vipers it has spawned.

So yet another red check against the Democrat Party, JUSTICE DEMOCRATS / DEMOCRATIC SOCIALISTS. They are is satanic. Ask the parents of their victims:

 # 3 HAS THE GLOBAL WARMING / CLIMATE CHANGE MOVEMENT COME TO STEAL ?

As we have seen in this book, money makes these movements go round...

In addition to the financial fraud, the Democrat Party, the JUSTICE DEMOCRATS / DEMOCRATIC SOCIALISTS are specialists in election fraud.

They steal elections by use of illegal alien voters during election day.

And after election day through fraudulent vote counts, fraudulent absentee ballots, and vote harvesting, they reverse the results of elections. The Global Warming movement is the movement of voter suppression.

Election fraud happened on a huge scale, for example in California in the November 2018 mid-term elections for the House of Representatives. And they tried to do it in Florida as well, but were stopped by an army of volunteer Republican attorneys and vote counting monitors.

More about stealing... how about the billions of dollars stolen by fake scientists writing fake reports for money?

How about stealing tax payers' money to stop cows from farting?

This isn't pocket change. It's many billions of taxpayer dollars every single year.

And as if that isn't enough, they seek to steal your property. In the name of SAVING THE PLANET from climate change. Massive taxation will reduce the people to poverty.

THEFT OF RIGHTS: They are working day and night to steal the rights of All Americans, enumerated in the Constitution of the United States. Free speech, right to bear arms, freedom of the press, freedom of religion, and on an on. In the name of their greater cause, the false religion of Global warming.

And of course, they are trying their best to steal the brains of American children. To turn them into mind numbed robots.

So on this matter of stealing, the Climate Change movement and its people receive the RED CHECK:

4 HAVE THEY COME TO DESTROY?

The main objective of the so-called Global Warming / Climate Change From Human Activity people has nothing to do with climate.

It has everything to do with POWER.

That is what they want.

The destruction of the American system of government, which guarantees liberty and equal opportunity to all its citizens.

And to put in its place, a Nazi-like totalitarian state.

The new thought leaders, piling in after Al Gore, might include Zach Exley and Becky Bond. As well as those who finance them with dark money. All apparently students of the master of Nazi propaganda, Minister of Propaganda, Joseph Goebbels.

"We enter congress in order to supply ourselves, in the arsenal of democracy, with its own weapons.

If democracy is so stupid as to give us free tickets and salaries for this bear's work, that is its affair.

We do not come as friends, nor even as neutrals. We come as enemies. As the wolf bursts into the flock, so we come. "

Destruction of rights, liberties, and democracy is the goal of the Democratic Party, JUSTICE DEMOCRATS / DEMOCRATIC SOCIALISTS, and the Democrat Media which supports them.

And on a family note, their open borders policies lead to the destruction of about 50,000 American families yearly. These families suffer death or injury, at the hands of illegal aliens who act in a commission of a felony crime.

The Democrat Party, JUSTICE DEMOCRATS / DEMOCRAT SOCIALISTS clearly qualify for a red check for destruction:

So yep, the Global Warming / Climate Change movement qualifies for the RED CHECK verification that it is Satanic.

In America, three strikes and your out. In the climate war, three red checks, and you're out. And we found four.

And if you don't believe this look at evil yet, tune in to the other chapters of this book which cover the genocidal and Nazi-like propaganda and power grabbing activities of the Global Warming / Climate Change From Human Activity movement.

It's not only demonic, it's a slam dunk to prove it.

CHAPTER 33

WHAT CAN WE DO NOW, AFTER READING OUR NEW BOOK

CAN WE STOP THE COMING ICE AGE?

GETTING READY FOR IT... A SUMMARY

As we have read, the Democrat Party, JUSTICE DEMOCRATS / DEMOCRATIC SOCIALISTS are working day and night to keep us from getting ready for the coming cooling period. They are a danger to survival of the US and must lose elections.

Ice ages are not unusual, and come and go. Because one is coming within a few years, we have to prepare.

Ice ages and warmer periods have been part of the earth's history for billions of years. They are the result of monumental forces far beyond human control or influence. Resulting basically from what is happening with the energy (heat) level of the sun. When it's very active, we have warmer periods. When it's in a minimum phase, we have cooler periods and ice ages.

About this, we can do nothing. So any idea of trying to modify a coming Ice Age is a total waste of energy. We can not stop it from occurring.

But we can get ready to some extent, for what is coming. And in getting ready for we can save countless lives. Both of humans, animals, and plants.

There are only about 35,000 sea otters left in this world. And we can save every one of them from freezing to death. This is do-able. We have the technology and resources to do it. And they are so cute, we just have to do it.

There are about 45,000 polar bears and we can save every one of them.

It will take effort to get water, power, food, shelter infrastructure ready for freezing cold weather. For not just animals but for us as well... But it can be done. Some countries have created such types of infrastructure already as part of their usual existence, for example northern Scandinavian countries. These countries are growing masses of vegetables during the winter when the temperature outdoors is below freezing.

These methods and technologies of the Scandinavian countries are good examples of what will happen in a serious ice age and what we can do about it to survive it with minimum damage.

SAVING THE SEA OTTERS...

The sea otters are about the size of a medium size dog, and live in water. Warm to moderately cold ocean water. But not frozen water. When the water freezes (and it will in an Ice Age) they will die.

The polar bears can survive in very cold water, which is partially frozen. But in ice ages the entire oceans freeze. From top to bottom. It means no food for them. And the severe cold temperatures would be too much for them.

So what can we do?

Massive projects to provide artificial environments, warmed by guess what.. fossil fuels.

Thanks to the new energy policies of President Trump, the United States is energy sufficient in oil and gas. We have gas and oil in abundance. This we can use to heat even vast areas of closed environments. Fossil fuels will enable us and the plants and animals to survive the coming Ice Age.

And in these artificially heated environments, we can not only save animals, we can grow food.

The future of our country is up to us, We the People. The obstacle between us and survival is the BIG LIE: Global Warming and Climate Change From Human Activity.

THE COMING POLITICAL WAR WE MUST WAGE TO PREPARE AND SURVIVE

In addition to the possible activities in the justice and judicial system mentioned above, we can take political and media steps as well.

2020 elections, as well as elections after that, are right around the corner, so it's good to start now. Because the totalitarian unhinged Democrat Party is working to prevent preparation for an Ice Age, political action against them of the most vigorous nature is a must.

For example, first, either by yourselves or in groups, go to the local Republican Party and join up. But join up with a plan to help them.

A plan to turn them into a more activist organization. One that is proactive not just reactive. Revitalizing the Republican Party is a means to success.

For example: **getting legal teams in every state ready for Democrat election fraud in 2020**. And be in place and ready to go BEFORE THE ELECTIONS.

When cheating is observed, take action immediately. Before and during the election. And afterwards.
If we wait until after the elections, it is much more difficult to correct the Democrat election fraud.

This fraud typically takes the form of:

- **Illegal alien voting.**
  Several universities have conducted studies which estimate that around 3 million illegal aliens vote in US national elections for president.

  This illegal alien vote fraud has to be a key target of ours to stop. It is a serious crime and should be referred to criminal justice authorities and the Federal and state election commissions along with documentation.

- **Campaign fraud by the Democrats.**
  For example, sending out false information about Republican candidates by the Democrat Party / Justice Democrats. And other sabotage and dirty tricks. Many of these are illegal, and should be reported to authorities along with documentation and crushed when they occur without delay.

- **Vote counting fraud by the Democrats.**
  The Democrat Party has shown itself to be not very good in math. And also not very good in the handling of ballots. Our ballots disappear, and new ones magically appear with Democrat votes.

  This is a serious crime and also has to be referred to the appropriate authorities along with documentation.

- **Vote harvesting**
  Democrats have passed laws in many states, allowing anyone to pick up votes and deliver them to the election polling places.

  The problem is that only Democrats do this, and because their "volunteers" are in possession of the ballots, the ballots can be thrown away if they are Republican votes, or can be changed if they are Republican votes.

These election frauds are serious crimes, and the new Republican election fraud teams have to be ready for it as well as the other forms of Democrat, JUSTICE DEMOCRAT / DEMOCRATIC SOCIALIST election fraud.

Thus it's vital that as many new Republicans as possible join your local party to revitalize it, and to make it activist.

While most party officials at the state level are doing their best, they need help from the grass roots.

At the national level it seems that sometimes, the party is asleep at the wheel. So help them also.

And take part in the electioneering itself.

Get involved in GOP party visits knocking on people's door. It's called canvassing. And an organized program which you do in groups (for safety) can be joined at your local Republican Party.

Read Republican books, magazine, publications. Listen to the best talk radio. Frequent conservative constitutional websites.

Talk to everyone you know. Hand out printed materials. Through your local party go door to door in organized campaigns **to inform the citizens of the choice when they vote:**

- Liberty and Freedom with Republicans, or totalitarian hell with Democrats. Democrats will choose your doctor, your politics, your education, your food, your everything.

- The choice is.. Open borders with the Democrats, or a safe secure country with secure national borders with the Republicans

- A strong economy where everyone does well and has an equal chance, with Republicans.

    A bankrupt country with mass unemployment, with Democrats

- Security at home and abroad, with Republicans.

- Rampant crime, drugs, robbery, murder, rape with Democrat policies. Policies which are intentional and meant to cause chaos and destruction.

And finally support Republican candidates who stand for the constitution, who support the president, who support a free and booming economy. We do not need any more RINOs (Republicans In Name Only). Mitt Romney for example, must be the subject of a recall election. Bring him back to Utah where he can raise chickens and be happy.

Work for Republican candidates that are really representing the party of Abraham Lincoln, Ronald Reagan, and President Trump. The president is fighting for us every day. Now we can fight to help him and the whole country.

That's the long and short of the choice.

Why mention political matters in a book about the Democrat Global Warming / Climate Change fraud?

Because the inconvenient truth is that the Democrat Party does not give a hoot about climate.

For the political Left, climate is only a tool. A way for them to gain tyrannical political power and enslave the people of this country.

If the climate programs of the Democrat Party, Justice Democrats / Democrat Socialists are actually law in the US, then individual freedom, family freedom, job and economic freedom is over.

The tyrannical unhinged Leftist elite will control everything you can do and will push you down into poverty such as exists in Venezuela, Cuba, North Korea.

While they of course live the high life.

Climate means nothing to the Democrat Party, JUSTICE DEMOCRATS / DEMOCRAT

SOCIALISTS and their billionaire backers. They want power.

POWER OVER YOU !

And the fake climate hysteria seems to them a good way to get it.

ABOUT THE INCONVENIENT ICE AGE

AND BECOMING A TRUE CLIMATE ACTIVIST

To end our book, dear readers, let's look at the real climate challenge. One which we discussed in great detail in our book.

It's been a great journey together. And now we can become the true climate activists.

Leading US and world scientists, and many have been presented in this book, have provided evidence that between this year and 2035, we are going to be hit with a new ice age.

It's not getting hotter. It's going to get a lot colder.

And are we ready for this as a nation? No.

And the reason why is that the damnable lie of the political Left, Global Warming / Climate Change. The BIG LIE is pushing us in exactly the wrong direction. Just as the Nazi Big Lie pushed Germany into national suicide.

Instead of getting ready for heat, we need to be getting ready for cold.

This means, as we have mentioned repeatedly, because it is so important, that among other infrastructure changes, hardening our water delivery systems against the cold.

Otherwise, water mains and pipes will freeze and burst. In large cities, people simply won't have any water. Try to imagine it.

The technology to achieve this resistance of the water delivery system to extreme cold is well known. But we are not doing it.

The electrical generation and delivery system also has to be hardened against the cold. Ice and snow will form on power lines and down they go. The freezing temperatures will damage power generation plants.

And because the transportation system will be frozen to a halt, almost all power plants will not be able to get fuel to produce electricity.

And specifically regarding transportation: truck engines will freeze, and the roads be filled with snow and ice that our cities are not capable to handle. Cars will suffer the same cold weather fate. They will freeze shut along with their frozen batteries. Delivery of gasoline to filling stations of course will also end because there are no clear roads or working vehicles to get it there.

For those living in apartment buildings, the ice age means: no water, no electricity for light or for running the elevators. Locks on doors will freeze shut.

An ice age is not pleasant, and we are doing absolutely nothing to get ready for it.

So as part of your Republican political activity, be a climate activist, working to get ready for the coming ICE AGE.

Join your local republican party and revolutionize it in to activism. The party can use your help. And together call out those on the Left who are pushing the United States into national suicide.

ICE AGE DENIERS

They must be crushed in politics, the judicial system, and in the Democrat Media Complex.

Use the data presented in this book. It's both accurate and effective.

Use this book as part of your toolbox. I am not trying to hawk my book, but use it yourselves and get copies of it to others. Particularly Republican decision-makers.

This book is a weapon for truth and liberty against the tyranny of the political Left. Use it. And may your work be blessed.

Let's save the polar bears and sea otters from freezing to death. Because that's what awaits them if we don't act against the ICE AGE DENIERS

The unhinged Democrat Party and its Democrat Media Complex, along with their climate mob, have declared war against We the People.

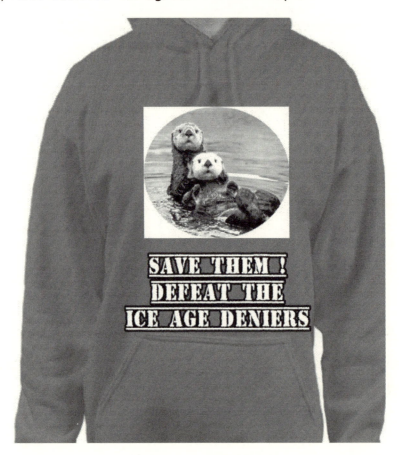

We have been dragged into a political, judicial, and media war.

It's going to be called **THE ICE AGE WAR FOR SURVIVAL**.

Put up the political, judicial, and media barricades and get ready to storm their dark castles of lies.
We can revolutionize the Republican party into an activist party for liberty, human rights, and life.

Yes life.

The Declaration if Independence makes it clear that all are born with inalienable rights for **life, liberty, and the pursuit of happiness.**

The Global Warming / Climate Change mob in congress, in the Democrat Media Complex, and on the streets is trying to take those away.

And because they are blocking preparations for the coming ICE AGE, they are hell-bent take life away along with everything else.

Join the Republican Party, or independently form political, judicial projects, and media groups.

Support the president in every way you can.

It's been a great voyage together, dear reader. And we can power up the first phase of our trip together with the inspiring words of Thomas Paine from the American War of Independence period. In a paraphrase for our times:

These are the times that try our souls.

The summer soldier and the sunshine patriot will, in this crisis, shrink from the service of their country;

But those that stand it now, deserve the love and thanks of every man and woman.

CHAPTER 34

THE SAD BUT PREDICABLE GENOCIDAL NATURE OF THE GLOBAL WARMING / CLIMATE CHANGE FROM HUMAN ACTIVITY MOVEMENT

AND WE BRING A CHALLENGE TO THEM

Yes dear reader, strangely enough, the people who are constantly telling us they want to "save the planet" are actually destroying it. And doing so happily.

They are destroying it in two ways:

First, we are heading for a disastrous new ice age. Yet the climate hysterics are sending us in exactly the wrong direction. Action is needed to build and protect infrastructures against the coming cold, not against heat. And these preparations are exactly opposite of one another.

Second, like Nazism, like Soviet Stalinist totalitarianism, Global Warming / Climate Change is based on the big lie. These always lead to disaster.

It's a standard tactic used by the political Left in America. And President Trump has been one of their victims.

Repeat the lie often enough, falsely slander and ridicule your opposition, then people will believe your big lie. And the Democrat Party. JUSTICE DEMOCRATS / DEMOCRATIC SOCIALISTS have put the Big Lie into turbo drive. That can't be good for the planet, can it?

Thirdly, and we have to face this with strength. Because similar to Nazism and Soviet Stalinism, the Global Warming / Climate Change hysterics have genocidal aspects.

Let's start with the cows. Cows and all other living beings create greenhouse gasses. These such as CO_2, are good for the planet. But of course the political and environmental Left has everything backwards.

AOC seems very worried about methane from cow farts. But she has failed to look in her biological mirror. AOC is not biologically a cow, but she as well at least occasionally emits methane in a similar way.

After a backlash against genocide of cows, and the prospect for the Global Warming troops of no hamburgers, a new cow theory has arisen. Now the hysterics claim they just want to change the cows' diet. Give them something different than commercial cow feed. Something more cool, you know, like veggie food, natural agriculture.

But cows of course get plenty of green grass and other of their natural food in the summer, when they can do outside. They still produce methane as part of their digestion.

And even if somehow commercial cow feed changes, the cows will still create methane. Then what?

AOC will be faced with over 100 million big living things, that no matter what, are going to fart methane. Then what? The answer is automatic from the climate hysterics... to save the planet the cows must be sacrificed.

Don't doubt it. The Democrat Party, the JUSTICE DEMOCRATS / DEMOCRATIC SOCIALISTS are quite OK with murdering millions of babies.

They love abortion. And support the abortion industry.

About 16 million African American babies have been aborted.

So do you think killing a few cows is going to stand in their way to "save the planet"?

To save the planet from a problem that is fake and does not exist.

Anyway we Americans are eating too many hamburgers, according to AOC. The country can eat peanut butter and bananas for breakfast, just like her. Full speed ahead on the cow genocide.

And they could go through with it. Both from a moral viewpoint, and because they have the money.

There are billions of dollars of radical Left secret dark money behind the Democrat Party, Justice Democrats.

Note that AOC's campaign was financed by dark money to the tune of apparently around two million dollars. And of course the political left can get can get even more from the taxpayers in the AOC totalitarian state.. to spend on exterminating the cows, or unborn babies, or what ever they want.

And that's how Democrat Party, Justice Democrats / Democratic Socialists and other totalitarians work. If they ever manage to get power, their cruelty knows no bounds.

They first isolate and then destroy one group. Take them out. Other groups don't resist because they hope they will be safe. In this instance it's the cows first. All in the name of saving the planet.

After that the Democrat Party, Justice Democrats / Democratic Socialists take out another group they don't like for one reason or another. Following the cows, it might be the Libertarians. Who knows.

With Stalin, he murdered 10 million Ukrainians by starvation, because they would not give up their independent farms. There are a lot fewer Libertarians in the US than that for the unhinged political Left to deal with.

For Stalin was just the starter, as he accumulated absolute tyrannical power, he then went on to eliminate another 60 million people.

THE NEXT STEP AFTER THE COWS

So dear readers, what comes next after the cows have been disposed of by the Democrat Party ?

The University of Illinois, the same pile of crud that provided the fake 97% of scientists study, have given us the answer:

White people.

Now, the author of this book is tan color. I'm a minority. That's good to know, to make the point that I'm not writing this section to protect myself. It's to protect ALL Americans.

To those who oppose the AOC, Omar, Talaib group, pigmentation doesn't matter one bit. Not to anyone with functioning brain cells anyway. We are all Americans.

But apparently skin pigmentation matters a lot to the fake University of Illinois scientists, the Democrat Party, Justice Democrats / Democratic Socialists in their Global Warming movement.

That's how fascism works. Isolate groups one by one. Take them out one group at a time.

Yep, after good bye to the cows, the next on the Global Warming / Climate Change chopping block will be, according to the new University of Illinois study... "White People".

The reason for this coming genocide from the tyrannical Left is provided for them by the new University of Illinois study. It's sort of like a *Mein Kamph* guidebook to genocide. Or a catchy tune from the Soviet Union which is most certainly popular in the halls of the department which wrote this amazing article about Global Warming and "White People".

The words of this Soviet song for the youth groups of the Soviet Union, have gotten through to them:

"Oh I want to be like Stalin.
In the home of the brave and free.
I want to be like Stalin.
He's the hero of my country!"

Any sane person, gets chills up their spine reading the University of Illinois study. Of which the authors, editors and university overseers are so blind, they do not see what they have done. The would be leaders of America are intellectually blind and emotionally unhinged. Not a good combination.

For the universities and their professors who are bought and paid for by the Global Warming money machine covered in detail in this book, this study and all their work, is all in a days work.

The dark money, plus their left wing lobotomies are all that is needed for such a study to be written.

They have been rendered so mentally disturbed, that they don't seem to get the fact that they themselves are "white". They have painted a bullseye on their own backs.

And the study they have written is trying to sew the seeds of their own destruction. Truly a Halloween scenario in which reality disappears and goblins become the norm.

Sensible street philosophers, locked out of the Leftist university ivory towers, state the obvious:

"The most stupid people in the United States are White, Leftist academics at universities and colleges. Either that, or their main hobby is suicide."

Anyway here is the journal article's argument for genocide from the Democrat Party academics. Put on your seat belt, and if you are wondering who would sponsor or publish this university piece of crap, you are not alone. Here it is:

JOURNAL OF INDUSTRIAL ECONOMY

Overcoming climate change adaptation barriers: A study on food-energy-water impacts of the average American diet by demographic group

https://onlinelibrary.wiley.com/doi/full/10.1111/jiec.12859

First of all, to dress up the heavy racism of this study, and its genocidal intent, the authors and researchers use the usual Democrat Halloween words to fool us. Instead of "white people", they begin the study by using the term "Caucasian".

Well they tried to put on a Halloween costume of civility, but as the study progresses, they just couldn't stop themselves from lapsing into the term "white people". I've never seen a white person. Well perhaps photos of albino individuals. But other than that rarity, no.

The theory of this nauseating waste of taxpayer and Leftist billionaire money is that "Caucasian populations (AKA "Whitey") "are disproportionately contributing to climate change through their eating habits, which uses up more food — and emits more greenhouse gases — than the typical diets of black and Latino communities."

Yes, the wicked Whites, are eating the planet to death according to the brilliant researchers. Is a Nobel prize for the authors in the offing?

But maybe not. If there was ever an example of a fake scientific paper which uses doubtful and unknowable "data" to prove an insane point, this is it. And unfortunately, in the Global Warming academic hysteria group, it's not unusual. It's possible to summarize and say that the Global Warming / Climate Change academic hysteria studies are scientific "crap".

The geniuses at the University of Illinois, appear to some experts as genetic mutations of Al Gore, made from a DNA sample from one of the 11 toilets in his mansion... these geniuses managed to grub up data from the National Institutes of Heath, on who eats what in America.

Oh my goodness. Like the National Institutes of Health know what what I (a tan colored person) or anyone else actually eats.

The pseudo scientific language of this study is enough to make any color person puke:

"Researchers tracked information from multiple databases to identify foods considered "environmentally intense" by requiring more precious resources such as water, land and energy to produce — and, as a result, releasing more greenhouse gases such as carbon dioxide through production and distribution.

Potatoes, beef, apples and milk are some of the worst offenders.

"The food pipeline — which includes its production, distribution and waste — contributes significantly to climate change through the production of greenhouse gases and requires significant amounts of water and land, which also has environmental effects,"

All this according to a student at the University of Illinois at Chicago, who helped author the study.

YIKES! It seems like the brilliant U. of Illinois researchers discovered that those "white deplorables" are gobbling up potatoes, apples, beef, milk and the other planet-destroying foods.

Actually according to the study, the difference in volumes of these foods eaten between "whites" and blacks and Latinos was very little. Everybody had a sort of similar diet. But who cares. It's apparently fun for them to write crazy Global Warming papers. Especially if you get funded a ton of money for it.

Still it seems the brilliant researchers missed a key point in their battle against Global Warming.

Aren't the whites actually heroes? According to the study, the "Whites" eat more beef.

Because isn't eating beef a way to kill cows?

The cows dreaded by the Global Warming movement? Cows which are destroying the planet with the methane they are expelling from their posteriors?

It seems then that eating beef should be a good thing for the planet. I'll volunteer to help with the propaganda, and those on the Global Warming Left who read this, feel free to use it. Yell it in your street protests from sea to sea:

EAT HAMBURGERS? REDUCE METHANE FROM COWS AND SAVE THE PLANET !

Missing that important dietary aspect of climate science and sociological science, was a foundational mistake in scientific method by the university team. But hey, rationality and reason never means anything to the Global Warming mob.

Then the study went on to glorify the diet of groups the Global Warming elite wish to spare from the coming Climate genocide. African Americans and Latinos.

"The EPA provided data on per capita food consumption rates for more than 500 foods groups, including water, plus estimates from the NIH on individual diets. Data showed that whites produced an average of 680 kilograms of the CO_2 each year, attributable to food and drink, whereas Latino individuals produced 640 kilograms, and blacks 600.

They also found the diets of white people required 328,000 liters of water on average per year. Latino Americans used just 307,000 liters, and blacks 311,800.

Both black and Latino individuals used more land per capita with 1,770 and 1,710 square meters per year, respectively, than white people with just 1,550.

Nevertheless, white people still made the greatest overall contribution to climate change."

Oh my goodness.....

First of all, like all genocidal ideologies, this so-called "study" is based on scientific fraud. The idea that a few cretans sitting in their University of Illinois cubbyholes could determine that any particular group in America actually puts into their mouth, is absurd.

The vegan movement, of course, for example is primarily "white", according to existing non-fake studies. So how does that fit in with the "meat" link to "whites".

In addition I probably would be counted among the Whites by these University of Illinois lunatics, but they would be wrong again. I'm genetically and naturally tan.

Yet in their haste to use only three categories of people... two being skin color ("white" and "Black"), and the other being cultural and geographic ("Hispanic"), the the researchers are making another academic catastrophic research mistake.

And as a "whitey", guess what... I eat no beef and few potatoes.

There goes the Nobel prize, for the University of Illinois.

Better science than theirs says simply, some people eat this and some people eat that. Pigmentation isn't a food magnet. But we really don't know who is eating what.

Pretending to have accurate data so as to project "climate damage", is insane and a lie. But what else could we expect from the Global Warming movement?

Another catastrophic data flaw is to say that a particular food creates more global warming. I know the Leftist theory that people who eat cows create more CO_2 from their diet. Yet that is really impossible to calculate for a multitude of reasons.

For example, the nutritional favorite of my friend AOC, which she recommends to her two million twitter followers (most of whom are perhaps Russian bots), is **peanut butter**.

She stated publicly her breakfast consists of bananas and peanut butter. Sounds like fun, but is it such a low $C0_2$ footprint? (Not that CO_2 matters, since the relation of CO_2 to climate is zero.)

Sooo, if we go through the whole process of farming, harvesting, cleaning, shelling, roasting, machine processing, and packaging peanut butter... the carbon footprint may be about the same as cow meat.

I don't know and I don't care who eats what because.... CO_2 has no relation to Global Warming / Climate Change at all.

When I originally wrote the part about my friend AOC's peanut butter, I did it as a joke. You know, to lighten up the book. But as it turns out, I had stumbled onto something which could, if picked up by the University of Illinois, earn them that Nobel Prize in chemistry.

Here is a summary of a real study on the carbon footprint of growing peanuts and making peanut butter. It's certainly more exciting and interesting than the University of Illinois motivation for genocide of "whites" article:

This peanut study investigates the environmental impact of peanut butter production in the USA. It identifies the following main environmental impact categories: natural gas production, disposal of municipal waste, disposal of coal byproducts, and pesticide use.

Natural gas production and use is associated with drying (buying point) and roasting. Disposal of municipal waste was associated with blanching, roasting, and processing. Disposal of coal byproducts is associated with processes on the farm. Pesticide impacts are primarily associated with farm operations.

However, interestingly, pesticide amounts reported to be also used at the sheller, blancher, and roaster are in quantities large enough to be used in the model.

Now this is really a thrilling study.

The study also investigates greenhouse gas emissions from peanut butter production. Although this was outside the study boundaries, it finds that the use and disposal phase cause the highest amounts of greenhouse gas emissions (for the electricity required to pump and heat the water for cleaning the dishes). Well, maybe not so thrilling.

But compared to the University of Illinois study, it's Nobel Prize quality. it's a slam dunk.

http://envormation.org/environmental-footprint-of-us-peanut-butter-production/

As a way to end this cheerful chapter on the genocidal nature of the Global warming & Climate Change Caused by Human Activity movement, here is an actual scientific fact they can't deny.

All human beings, regardless of pigmentation or culture, exhale CO2.

24 hours a day, seven days a week. We are little CO2 factories.

Thus we have the incredible situation in which the apparent new leaders of the Global Warming movement, Al Frostbite Gore, AOC, and Mazie Hirono... along with the brains behind the Justice Democrats and Democratic Socialists, Zach Exley and Becky Bond... well how should I say this?

I apologize if this is indelicate, for the sake of science I write it to achieve a brighter future, these Global Warming hysterics not only exhale the dreaded CO2, but they fart methane as well !

In fact all those involved in writing, editing, and even thos paying tons of money for this insane paper from the University of Illinois... all of them exhale CO2 and fart methane.

This shocking news has ramifications for these Global Warming hysterics.

Thus ladies and gentlemen and transgender readers, or the new no-sex category... this issue of Democrat leaders exhaling CO2 and farting methane, now becomes a plea to save the planet.

Those who are part of the Global Warming movement, no matter where and who you are... if you really believe the crazy idea that CO2 is killing the planet, then take the next step. The virtuous step. The brave step: stop breathing. Stop passing methane gas.

Take a hit for the team!

Come on, you can do it !

To the University of Illinois social justice and climate warriors who wrote the amazing racial genocide motivation study, why not be the first in line.

In addition to the genius authors of the diet genocide study, how about all their so-called professors who approved it at what we must now refer to jokingly as the "University" of Illinois.

(And how can we forget the other major U. of Illinois paper found in the beginning our book, which created the 97% of scientists lie?)

And why not the groups financing this genocidal paper also taking a hit for the planet:

The Institute for Environmental Science and Policy of the University of Illinois at Chicago and the Diversity Science, Technology, Engineering, and Mathematics Fellowship from Bayer-Monsanto.

And the Editor of course.

And of course AOC, Frostbite Gore, Daisy Hirono, and all the Democrat Party frauds in congress who support the Green New Deal… All the billionaires who provide funds for this "crap": come-on gang! Get involved personally. Take a hit for the team.

Take a hit for the planet.

Remember and be inspired: AOC has demanded of the protest mobs when she encouraged them as they sat on their behinds on the floor, in Nancy Pelosi's office protesting: "put your body on the line".

But praise the Lord, no one has to stop breathing. Or stop expelling methane.

Because CO2 and Methane have no effect on climate.

This truth is a truth of life on earth. And it allows of the above mentioned people, to slink off into the shadows, unwilling to make actual personal sacrifice.

While the authors of this hideous academic paper sit and worry, the rest of the country is heading off into the kitchen for a meal of the forbidden foods potatoes, apples beef, milk (in the form of yogurt or kfir for the more Woke people among them).

And some cookies to boot.

I mention this because this study from the University of Illinois, and the whole Global Warming lie is like sewerage water running all over the place. No matter how much it stinks, you can't ignore it. It's rotten to the core.

But happily, the nation is going to eat, meat and potatoes and apples and milk despite this study. Because we know that CO2 and Methane has zero effect on climate.

And finally as a nutritional matter, it's useful for the future of the country to mention something about AOC's peanut butter. Science has to be thorough… I myself like peanut butter.

As a kid I a lot of it. But it does contain nasty little things called Aflatoxins.

Those will destroy the planet long before CO2 and cow farts dreaded by the Global Warming hysterics do.

And if the anyone mentioned in this chapter criticize me of anything, well, that's RACISM. I'm tan color, don't you know.

I'm also an accredited member of the press. Any such criticism of this chapter, or in fact anything in this book, by the unhinged Justice Democrat types or Democrat Party, is understood to be an attack on Freedom of the Press. An Racism as well. I learned this from the Leftists.

CHAPTER 35

THE GREEN NEW DEAL...

AND THE PEOPLE BEHIND IT, ARE A THREAT TO NATIONAL SECURITY OF THE UNITED STATES

AND THIS IS NOT A JOKE. THEY REALLY ARE

WHO THEY ARE AND WHAT THEY ARE DOING

In the previous chapter, we explored the genocidal aspects of the Global Warming – Climate Change – Green New Deal crowd.

If you were, dear reader, skeptical that the Green New Deal, and the new type Global Warming movement has a relationship to Nazi methods, tactics, ideas and objectives.... just wait until you read this chapter. You will be convinced.

Note that no person and no organization mentioned in the book is accused of committing a crime. All Americans are presumed innocent until proven guilty in a court of law.

Get ready for an interesting ride. Buckle up.

You won't be skeptical any more. A massive Neo-Nazi movement, well financed, is active and in plain sight for those who look carefully.

Let's take a look, shall we?

THE JUSTICE DEMOCRATS PARTY AS A NATIONAL SECURITY THREAT

AND AS A NAZI IDEOLOGY MOVEMENT

THE FOUNDERS AND THE BRAINS BEHIND IT

Financial founding in its very roots and beginnings, came from George Soros funded organizations. That's not a great start.

Soros' own home country, Hungary has booted his organizations out, because their intent has been determined to destabilize and overthrow orderly, democratic republic government. According to the Hungarian elected officials.

Soros himself is a self-admitted Nazi collaborator during World War II. He has stated that as a teenager, he assisted the occupying German forces to identify and take the property of wealthy Hungarian Jews.

Although born into a Jewish family himself, Soros became a sort of adoptive part of a Hungarian official and family that worked with the Nazi occupiers in the extermination of Jews in the capital and theft of their property.

The present funding of the Justice Democrats / Democratic Socialists from Left wing billionaires is unknown. Perhaps the organizations funded by Mr. Soros are out of the picture. It's not known. But at the start it was nurtured by Soros-funded organizations which nurtured those who are leaders today.

Dear reader, why mention all this in a book about climate? Because in this chapter we focus on the Global Warming fraud and the Green New Deal push to destabilize the American economy, government, liberty and the constitution. It's all inter-related. In fact as we said before, the Global Warming / Climate Change Caused By Human Activity movements have nothing to do with climate. It's all about power.

The so-called "brains" of the JUSTICE DEMOCRAT party, are apparently good old Zachary Exley and his side-kick Becky Bond.

Exley was nurtured and supported by Soros funded organizations since 2004. For how long we do not know. These organizations include but are not limited to The **Open Society Institute**. There Exley had his indoctrination and start.

The language of the Left is amazing. The meaning of "Open Society" is "Totalitarian Oppression"

That the funding of JUSTICE DEMOCRATS today comes only from small donations, is very likely a fraud.

Directly, and indirectly through various PACs and Left wing billionaire "environmental" and social-political organizations, input of various types could amount to a very large part of their budget. It's unknown.

But that's for investigators. I'm a climatologist. If any readers could look into that funding matter, please do so. And publish it.

NATIONAL SECURITY RISKS FROM THE JUSTICE DEMOCRAT PARTY AND THE GREEN NEW DEAL

There is no reason to look very far to see this. It's clear and right in front us:

FIRST THE PEOPLE THEMSELVES:

The association and affiliation of at least two of the JUSTICE DEMOCRATS now in congress, are with terrorist supporting organizations. Some of them actual terrorist organizations, such as Hamas and the Muslim Brotherhood.

The goal of these people, the whole bunch of them, and they are doing it, is to disrupt the American social compact. And to spread conflict among various groups in America. To open American borders as a super-highway for illegal immigrants. Who injure and kill Americans in massive numbers, and place an unbearable burden on the economy.

America is more than any place else on earth perhaps, somewhere where its people of all races and religions basically can live together in harmony. THE CLIMATE CHANGE MOB IS OUT TO CHANGE ALL THAT.

The so-called JUSTICE DEMOCRATS are part of a racist campaign against Caucasian ("white") men, women, and children of the United States. As well as against Christians and Jews and political conservatives of all colors and gender orientation.

Such activities of the so-called JUSTICE DEMOCRATS, are of course a threat to national security.

As part of the disruption, they are promoting racism and antisemitism. Things they believe in deeply. As mentioned, this is actually a Neo-Nazi ideology party.

National security threats can come from within as well as without. The ideology and acts, such as they are promoting, are just as dangerous as threats from enemies outside the US.

President Reagan said, that the US can not be defeated and destroyed from without. But only from within. The so-called JUSTICE DEMOCRAT PARTY fits the description nicely according to experts.

The so-called Justice Democrats / Democratic Socialist follow the instructions and methods of the Nazi party of Germany during the 1930's used to overthrow democracy in Germany.

Exley, Bond and evidently the dark money behind them, follow in an amazingly close way the power grab and propaganda actions and teachings of the Nazi political party of the 1930s in Germany, the National Socialist Party.

Exley and Bond proclaim themselves as geniuses in communication, messaging and political communication.

But they are simply following the Nazi playbook on propaganda and how to overthrow free and fair republic elections.

THE GREEN NEW DEAL is basically Nazi type propaganda. With a totalitarian objective.

Here is what their master teacher, Joseph Goebbles (Minister of Propaganda for the Nazi government) said about how the Nazis used the German congress, an institution of democracy, to destroy democracy.

The Justice Democrats are on the same page and have been since the viper's nest was hatched in 2004:

"We enter congress in order to supply ourselves, in the arsenal of democracy, with its own weapons.

If democracy is so stupid as to give us free tickets and salaries for this bear's work, then that is its own affair.

We do not come as friends, nor even as neutrals.

We come as enemies.

As the wolf bursts into the flock, so we come."

Just as a reminder note to the giddy Justice Democrats over success in getting 7 of their candidates into the US House of Representatives by lies and deceptions about who they are and what their objectives are...

here is how their model and guide, Goebbels ended up. And in his own words:

"I'm so despondent about everything. Everything I try goes totally wrong. There's no escape from this hole here. I feel drained. So far, I still haven't found a real purpose in life. Sometimes, I'm afraid to get out of bed in the morning. There's nothing to get up for."

As Russian troops closed in on Berlin, Goebbels committed suicide by cyanide capsules.

Here is a photo of the Goebbels family, with their beautiful children taken some years before the end of the war:

Shortly after this, Goebbels killed and wife and six beautiful children along with him.

Take note JUSTICE DEMOCRAT leaders, how your idol and teacher ended up.

Any such advice will fall on deaf ears, because the members in congress, members on the street, and brains behind the JUSTICE PARTY have themselves fallen for the BIG LIE. They not only tell it, they have fallen for it.

THE GREEN NEW DEAL IS NOT ONLY AN INSANE POLICY, IT IS A POLICY MADE BY INSANE PEOPLE. NOT LIVING IN REALITY.

As their master Goebbels said:

"Propaganda works best when those who are being manipulated are confident they are acting on their own free will."

Thus the leaders and brains behind the JUSTICE DEMOCRATS while following the lead of Joseph Goebbels, have unknowingly have swallowed their own BIG LIE themselves.

THE GREEN NEW DEAL ACTUALLY AS NOTHING TO DO ABOUT CLIMATE. THESE PEOPLE DON'T GIVE A HOOT ABCOUT CLIMATE.

IT'S A PLAN TO GAIN POWER, AND DESTROY THE FREEDOMS OF ALL AMERICANS UNDER THE CONSTITUTION.

The idea is: if the emergency is huge enough, and if it is dangerous enough, then they can shred the constitution. Anything to protect the world against Climate Change From Human Activity and Global Warming.

Two things, which as you know, dear readers, do not exist.

It's the Nazi way. The Justice Democrats.. **JD Take a look at their logo above. Does it look**

familiar?

A red background has been added to make things clear. Let's compare this to the National Socialist (NAZI) "logo".

The components of the "**JD**" in their logo are clearly elements of the Nazi Swastika symbol. Not of course accidental, not of course coincidental. It's planned by their leaders, their dark money billionaire financiers. Everything about their movement, their dark money, their brains, and including some of their new members of congress, screams the words:

TYRANNY

THE BIG LIE

And what is the big lie? Well, one of the three **BIG LIES** is... Global Warming and Climate Change Caused by Human Behavior.

THIS LIE HAS NOTHING TO DO WITH CLIMATE AND EVERYTHING TO DO WITH GRABBING TYRANNICAL POWER. TO TAKE THE RIGHTS AND FREEDOMS OF EVERY AMERICAN

GLOBAL WARMING / CLIMATE CHANGE IS SIMPLY THAT. A BIG LIE, MODELED ON THE PROPAGANDA METHODS OF NAZI GERMANY. AND IN FACT THE POLITICAL POWER GRABBING METHODS OF NAZI GERMANY.

SAD TO SAY, BUT LET'S TELL IT LIKE IT IS.

But we must admit. The Global Warming / Climate Change movement of the Justice Democrats is politically correct.

They do practice inclusiveness.

How are they inclusive? They include not only German Nazi propaganda and power grab methods, they include also Russian Soviet methods. Don't believe me? Look at this.

The above photo is their very own website and election propaganda photo.

Seems to remind us of something ????

YIKES !

NOW, HOPEFULLY THIS BOOK

DEAR READER WILL HELP YOU. WILL HELP YOU PUSH BACK AGAINS THE LIES. THAT THIS BOOK WILL BE A ONE PUNCH KNOCKOUT ON BEHALF OF LIFE, LIBERTY AND THE PURSUIT OF HAPPINESS.

CHAPTER 36

COUNTDOWN

WHEN THE ICE AGE IS GOING TO BEGIN... AND HOW WE CAN SAVE THE SEA OTTERS. AND THE REST OF US

While the Democrat Party and the Justice Democrat vipers they have spawned, work to destroy the country, nature is doing what it wants.

THE INCONVENIENT
ICE AGE

How is this projection of a cooling period different from the absurd and fraudulent hysteria of Al Frostbite Gore and the whole Global Warming movement?

First, in getting ready for an ice age, no one is grabbing power or reducing liberties and freedoms guaranteed under the constitution. The first and second amendments and all the amendments in the Bill of Rights are in fact the way to survive an ice age by getting ready for it through constitutional processes. And then maintaining social cooperation during it.

Secondly, we're not basing the dates on fraudulent data. And on absurd computer models. The solar computer model we use has proven nearly 100% accurate.

We have the three components which enable an accurate forecast that an ice age is coming soon, and when:

- NASA atmospheric heat loss into space, the highest in recorded history
- Oceanic currents are at a low level not seen since an Ice Age 1,500 years ago
- The world's most accurate solar cycle model, sets the dates

All these have been covered previously in our book.

According to these three markers, we don't have much time left to get ready.

Getting, water, electricity, transportation and food systems ready will take at least ten years and we don't have much more than that.

We have in fact for about 200 years been coming out of the last ice age. Now we are starting a slow dip into the next which will accelerate into an uncontrolled slide into the freezer.

It may happen sooner than this timetable. One group of scientists indicates it could start in 2019 or 2020. And the other sometime around 2035. But all agree we're sledding down into the ice age.

WHAT ABOUT THE CUTE LITTLE SEA OTTERS?

If it gets too cold, they can't get food in the frozen water. Their food supplies (mainly shelled crabs and oysters and so on who live at the shallow bottom of the ocean shoreline), will be gone. They also won't be able to withstand the very cold frozen water environment.

Frostbite Gore, AOC, the Democrat Party, the JUSTICE DEMOCRATS, and the Democrat Media Complex, pushing the Climate Change and Global Warming lies, have in fact been waging a jihad against the Sea Otters.

So if you've been brainwashed and have been a supporter of the fake Global Warming / Climate change, you have unknowingly been a jihadist against the sea otters.

It's not too late. Join, support and vote for candidates that crush the Green New Deal, which is in fact the Green New Steal. Or even worse, it's a recipe for the Green New Death.

That means supporting Republican candidates who commit to fighting against tyranny of the political Left. The Democrat Party and their little vipers the Justice Democrats / Democratic Socialists.

This is your chance to save the sea otters. Be happy you know the truth. Knowing the truth you get out can get out of the chains which the Global Warming false priests have put you.

Step forward, join, support,
vote for Republican candidates who can get our country for the next and around-the-corner Ice Age.

If you've been fooled by the Global Warming / Climate Change fraud, now you can change **and go from zero to hero**

AMAZING GRACE

And dear reader, as I now complete this book, I realized that there is something important I did not spend enough time on.

What reminded me, was when I added to the section about wishing that Alexandria would stop telling little white lies (and whoppers too) become a Republican. That she could read this book, and also listen to Judy Collins singing Amazing Grace.

So that brought up the question: in the coming Ice Age as the electricity cuts out, and batteries run out on our computers and smart phones and Ipads, what is the last song or music we will want to listen to?

One might be **Amazing Grace**, with its inspiring beautiful words:

"Amazing Grace, how sweet the sounds that saved a wretch like me.
I once was lost but now I'm found. Was blind but now I see.

'Twas grace that taught my heart to fear,
And grace my fears relieved.
How precious did that grace appear
The hour I first believed.

Through many dangers, toils and snares
I have already come;

'Tis grace hath brought me safe thus far
And grace will lead me home."

All the yelling, screaming lies of the political Left about Global Warming will end in freezing nothing.

Those who have fought against it, and fought for the God Given Inalienable Rights of the American people in the Constitution, will have something to hold on to.

And as one of the greatest sports people in American history, Joe Louis, said about America during the Second World War:

"We are going to win, because the God is on our side."

Yes. we are on God's side. Fight the good fight, and hold on. We will win.

DISCLAIMER

The information in this book is true and accurate to the best of our knowledge at the present time. It is in fact impossible to know when the coming ice age will occur. It could be in weeks. It could be in a few months. It could be several years. No one can say exactly, just as we can't exactly predict the weather the next week or month exactly.

But because the sun entering a very cool phase, a solar minimum, and because of the oceans current-flow data, and from the NASA data showing that the earth is losing tremendous heat through the atmosphere and stratosphere into space... the future according to this data is clear.

A new ice age is going to happen and probably relatively soon. But exactly when we will not know until we are in it.

This book is not an attempt to panic anyone. Or to incite anyone. Quite the opposite.

We need to be calm and get ready by hardening our energy, food, transportation, and water infrastructure against freezing. We need to get back to a civil and cooperative society. And act politically, judicially, and in the media against the Democrat Party, JUSTICE DEMOCRAT / DEMOCRAT SOCIALIST policies of division, hatred, chaos and lies.

Now.

No one in this book, no organization named in this book, is said to have committed a crime. In the US system of justice, all are assumed innocent until proven guilty in a court of law.

The information in this book about said individuals and organizations are accurate to the best of our knowledge at the present time, and are based on multiple sources.

This book is written by an accredited member of the press, under the First Amendment to the Constitution. Freedom of the Press and Free Speech.

has helped him understand what an Ice Age actually will be like.

He is an accredited member of the constitutional conservative press. And writes this book under the First Amendment: Freedom of the Press, and Freedom of Speech.

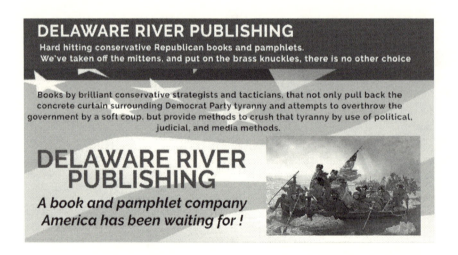

delawareriver@usa.com

Made in the USA
Coppell, TX
29 December 2022